YOUR STATISTICAL CONSULTANT

D0047722

RAE R. NEWTON • KJELL ERIK RUDESTAM

YOUR STATISTICAL CONSULTANT

ANSWERS

TO

YOUR

DATA

ANALYSIS

QUESTIONS

SAGE Publications
International Educational and Professional Publisher
Thousand Oaks London New Delhi

For information:

SAGE Publications, Inc.
2455 Teller Road
Thousand Oaks, California 91320
E-mail: order@sagepub.com

SAGE Publications Ltd.
6 Bonhill Street
London EC2A 4PU
United Kingdom

SAGE Publications India Pvt. Ltd.
M-32 Market
Greater Kailash I
New Delhi 110 048 India

Printed in the United States of America

Library of Congress Cataloging-in-Publication Data

Newton, Rae R.
    Your statistical consultant: Answers to your data analysis questions /
by Rae R. Newton and Kjell Erik Rudestam.
        p.   cm.
    Includes bibliographical references and index.
    ISBN 0-8039-5822-6 (cloth: acid-free paper)
    ISBN 0-8039-5823-4 (paperback: acid-free paper)
    1. Social sciences—Statistical methods.   2. Statistics.
    I. Rudestam, Kjell Erik.   II. Title.
    HA29 .N458   1999
    001.4′22—ddc21   98-58135

99 00 01 02 03 10 9 8 7 6 5 4 3 2 1

Acquiring Editor:	C. Deborah Laughton
Editorial Assistant:	Eileen Carr
Production Editor:	Diana E. Axelsen
Editorial Assistant:	Nevair Kabakian
Typesetter/Designer:	Danielle Dillahunt
Cover Designer:	Candice Harman

Contents

PART II: THE LOGIC OF STATISTICAL ANALYSIS
Issues Regarding the Nature of Statistics and Statisical Tests

PART III: ISSUES RELATED TO VARIABLES
AND THEIR DISTRIBUTIONS

PART IV: UNDERSTANDING THE BIG TWO
Major Questions About Analysis of Variance
and Regression Analysis

List of Tables and Figures

Preface

    Your Statistical Consultant is based on the questions that students and colleagues have in their attempts to apply statistical theory to real-life research problems. Both of us have trained and supervised graduate students, in our respective fields of sociology and psychology, for many years. Our students have been required to take courses in basic and even advanced statistics, usually from competent instructors using standard, popular texts. Over time, we have noticed a certain predictability to the kinds of statistical questions our students have asked as they engage in the pragmatic and challenging tasks of designing their studies and analyzing their data. Probably the most frequently heard question is, "how many subjects do I need?" the implication of the question seems to be, the fewer, the better! Sometimes we have had ready answers to these questions, particularly in those instances where we have heard them voiced time and again, giving us the opportunity to conduct a little research of our own. At other times, we have had to consult with a true expert in the field of statistics to be maximally helpful to our students. In some instances, we have found that there is no uniform agreement to the answer, so that our advice centers on how to justify the choice of a particular approach.

    Several of these questions are rooted in controversy, and what the student is requesting is practical guidance in completing a successful study. In this book, we discuss the issues that pertain to a range of statistical questions and controversies, reveal divergent perspectives on these issues, and offer practical advice and examples for conducting statistical analyses that reflect our interpretation of the consensual wisdom of the field. In so doing, we have provided a guidebook that should be useful to graduate students in the social sciences as well as researchers and professionals who want to stay current with the practice of statistics, who have questions regarding how to approach a specific statistical problem, or who are seeking consultation on how to approach statistical issues in the classroom setting.

Your Statistical Consultant is neither a traditional statistics text nor an advanced text detailing the study of a specific statistical technique. Rather, we view this book as a compendium of statistical knowledge—some theoretical, some applied—that addresses those questions most frequently asked by students and colleagues struggling with statistical analyses. The book may be used as a personal guide, as a supplement to a basic text on statistics, or as a reference for addressing specific issues in a research or teaching context.

We have attempted to orient *Your Statistical Consultant* toward distinct questions in a "How do I?" or "When should I?" format, understanding that the answers to these questions may require a refresher or reorientation to specific elements of statistical theory. We found that to provide a context for posing our questions and answers, it became necessary to explain fundamental statistical theories, as well as contemporary developments that are altering the face of the field. We have also had to rethink our own training, particularly in the light of the application and misapplication of statistical significance testing. We cover these elements in a nonmathematical manner in two chapters on the logic of statistical testing and one chapter on the meaning of the assumptions underlying statistical inference.

We have many individuals to thank for their wisdom and support in helping turn this idea into a finished project. First is our continually supportive editor at Sage Publications, C. Deborah Laughton. Next are the reviewers of early drafts of the book: Keith Haddock, University of Missouri; Jane Monroe, Teachers College, Columbia University; John C. Ory, University of Illinois–Urbana; Catherine Hackett Renner, West Chester University, Pennsylvania; Mary Colette Smith, University of Alabama at Birmingham; Anita Van Brackle, Kennesaw State University, Georgia; W. Paul Vogt, Illinois State University; and John Delane Williams, University of North Dakota. Third are our compulsively helpful copy editor, A. J. Sobczak, and our production editor, Diana Axelsen. Fourth are the many students who challenged us to become more knowledgeable statisticians by seeking our guidance in answering their own questions. Fifth are the many "true" statisticians who are cited throughout the book. We provide extensive references to their work. Finally, there are our wives, Katherine and Jan, and our children, Ryan, Kallie, Kirsten, and Monica, who offered ongoing emotional support and gave us a wide berth to complete this project.

Introduction

● 〜 Many excellent texts help make the dense and oftentimes impenetrable jungle of statistics comprehensible to and manageable by the social and behavioral science researcher. It is one thing, however, to understand statistical calculations and quite another to develop a conceptual understanding that facilitates the appropriate use of statistics in practice. Moreover, it is not uncommon for experts in the field to disagree with one another about statistical choices and strategies. In our experience, most professional academicians and researchers, not to mention undergraduate and graduate students, manage to identify a colleague with statistical expertise whom they can seek out routinely to answer their questions and guide them through the more vexing issues.

We invite you to think of this book as your statistical consultant, the colleague down the hall who has fielded all those inquiries through the years that deal with areas of concern regarding the logic of statistics and the analysis of quantitative data. In the absence of such a wise colleague on an as-needed basis, we have created this resource compendium to introduce, describe, explain, and make recommendations regarding predictably challenging statistical issues. This is not a book that has to be read from start to finish, although there is a certain organizational flow. Some readers may wish to begin at the beginning, while others may choose to turn directly to sections corresponding to individual concerns and questions. We have provided just such a list of questions at the beginning of each chapter. Areas of concern can be addressed quickly by turning to the page referenced. We assume only that you have a modest understanding of fundamental statistical concepts and procedures. The content in the book ranges from quite basic to quite advanced and sophisticated, so it should be responsive to a relatively wide range of inquiries and levels of expertise.

Your Statistical Consultant is organized into 11 chapters and four parts. Part I begins with a chapter on preparing data for analysis by computer. It is unlikely in this day and age that you plan to analyze your data without the assistance of a personal or mainframe computer. The data will need to be coded in some intelligible fashion for a statistical software program to manipulate the data correctly. This chapter presents a framework for coding data and then entering it appropriately into the computer for analysis. Chapter 2 deals with the first steps in statistical analysis, particularly with the visual presentation of univariate (one variable) and bivariate (two variables) data. Statistics can obscure the results of a research study as easily as they can reveal the meaning of the data. Consequently, it is generally a good idea to begin with an exploratory approach to fully understand the characteristics of your data. This chapter presents suggestions for developing graphs and tables, and it provides illustrations of useful figures such as stem-and-leaf diagrams, crosstabulations, and scatterplots.

Part II of the book explores the logic of statistical analysis. Chapter 3 begins with the foundational bedrock of inferential statistics by describing the logic of inference. After a thorough description of statistical significance testing, we continue in Chapter 4 to discuss the meaning of statistical power, effect sizes, and confidence intervals, and include some concrete advice on the topic of how many subjects you need for any particular study. We also include a discussion of contemporary challenges to traditional significance testing and the implications of these criticisms for the practitioner. Chapter 5 is a prelude to the selection of the appropriate test for a particular analysis. It deals with the assumptions underlying statistical tests. We attempt to provide the reader with a working knowledge of the meaning of the various assumptions most commonly associated with statistical testing and a rationale for their existence. Examples include the normality of distributions and that ten-dollar word found in every basic statistics text, "homoscedasticity." the concluding chapter in this part of the book, Chapter 6, offers explicit guidance on selecting the appropriate test to match a particular research question to a particular set of data. The chapter suggests that you follow a framework based on asking specific questions about the form of the data and the purposes of the analysis.

The remaining sections of the book address more focused questions about statistical application. Part III deals with issues related to variables and their distributions. One possible outcome of a quick examination of data is the realization that some of the data aren't available! This particular outcome is frustrating and leads to the question, "What if my data have missing values?" This is the first topic of Chapter 7. Viable answers range from ignoring the missing values to imputing new values from the existing data. Chapter 7 also tackles the problem of outliers, those numerical values that stray far from the bulk of the data. The chapter offers some guidance on how to gauge the impact of outliers and whether to keep them or delete them.

Chapter 8 asks and attempts to answer three questions concerning the understanding and appropriate use of variables and scales. We consider the possibility that your data may not reflect an interval or ratio scale, assumed by parametric statistics. How serious is this concern, and what are the alternatives? We also introduce the concept of "dummy" or "binary" variables, commonly utilized, but poorly understood, by most of us. We illustrate some ideas for enhancing statistical analyses with the use of dummy coding. Finally, Chapter 8 asks the question, "When, if ever, should I dichotomize a continuous variable?" the answer gets at the heart of the distinction between correlational and experimental problems and has important implications for managing power.

The questions in Part IV deal with the two most popular multivariate approaches to data analysis, analysis of variance (Chapter 9) and multiple regression analysis (Chapter 10). Both chapters open with a description and explanation of the "nuts and bolts" of the technique. Chapter 9 considers the proper interpretation of interaction effects. It turns out that interaction effects are often misunderstood and misinterpreted. We attempt to set the record straight. We then address the deceptively simple question, "How do I analyze the pretest-posttest control group design?" This, of course, directly addresses the question of how to measure change. Statisticians have devised a number of alternative procedures to assess change from pretest to posttest, among them gain analyses, repeated measures designs, analyses of covariance, and time-series designs. A thoughtful answer to this question must deal effectively with the need to control for preexisting differences between treatment and control groups. Another issue often connected with the analysis of variance is the differences between planned and post hoc comparisons. The chapter also describes the advantages of planned contrasts and explicates how to conduct them. We then address the common problem of increased error rates resulting from conducting multiple statistical comparisons. Chapter 9 contains some suggestions for how to keep the error rate low without sacrificing too much in the way of statistical power. Finally, Chapter 9 explains the proper use of multivariate analysis of variance (MANOVA) and the difference between conducting individual analyses of variance and a single MANOVA in instances where you have more than one dependent variable.

In Chapter 10, we confront typical problems encountered with multiple regression analysis, a second set of very popular methods of statistical analysis. The section opens with a description and explanation of the multiple regression procedure and moves toward a consideration of four frequently asked questions. The first question raises once more, but in a new context, that most common of concerns, "How many subjects do I need?", and offers a concrete reply. The next section explains the difference among regression techniques, including regular regression and stepwise and hierarchical approaches, along with the proper use of each. A very common source of confusion about regression concerns the meaning of the various coefficients or "weights" produced by regression analysis. The next section explains regression coefficients and how to interpret them. This should enable you to understand related terms, such as beta weights, R^2, and adjusted R^2, which all appear as part of the output of regression analyses. Finally, Chapter 10 offers advice on how to deal with interaction terms in multiple regression. This involves setting them up and interpreting them correctly.

The final chapter of the book, Chapter 11, is titled "The Bigger Picture." Here we consider macro issues that go beyond the consideration of any single statistical technique. The first section provides an explanation of how seemingly different statistical techniques are conceptually and mathematically related to one another. We hope this will take some of the mystery out of seeing statistics within a larger context. We then introduce meta-analysis, an increasingly prominent method of analysis. Meta-analysis is an approach to statistical analysis by which results from individual studies on a similar topic are pulled together and summarized to obtain a richer perspective on a body of work. The concluding section offers our final thoughts on summarizing, discussing, and making decisions based on statistical analysis.

We hope that you, the reader, find this book of value and that it assumes a prominent space in your library of statistical reference tools.

PART one

Getting Started With Statistical Analysis

Researchers typically collect data under the assumption that a computer will be used to analyze it. At least two important steps lie between the collection of data and its computer-based analysis using advanced statistical methods. One must first properly "prepare" the data for entry into a computer file or **database**, and once the data are correctly entered, one should examine the data distributions of each variable. There are many perils and pitfalls that can derail even an experienced researcher at these critical and necessary steps. To put it bluntly, if you err early, all later analyses, no matter how sophisticated, could be meaningless.

Part I is designed to alert the reader to these important early steps in data entry and analysis. The first chapter deals with the question of getting the data into a format that can be accessed by most statistical computer programs. This process is called **coding**. We discuss how to code both closed- and open-ended survey data. We highlight the importance of creating a dictionary of the data format, or **codebook**, and give an example of a completed codebook. The second chapter introduces data exploration or **exploratory data analysis (EDA)**. We believe that prior to any statistical "testing," one should examine the structure of the data for both data entry and coding errors. The researcher also should develop a mental image of the distributions of the variables composing the database. This is accomplished by producing both visual and quantitative summaries of the data distributions. We address this process in Chapter 2.

The first part of this guidebook should be read by the beginning researcher who has not had experience with data entry and wishes to understand the process of preparing data for entry into a statistical program, either directly or from another program such as a spreadsheet or word processing program.

CHAPTER ONE

How Do I Prepare Data for Statistical Analysis?

What is . . . ? What are . . . ?

How do I . . . ? When should or shouldn't I . . . ?

Code data for computer analysis?

Code data for statistical analysis?

Code data into smaller categories?

Code missing values?

Code open-ended questions?

Code open-ended response choices?

Create a codebook?

Create missing value codes?

Define missing values?

Enter data into a statistical program?

Handle blanks or blank spaces in data files?

Handle missing data?

Handle zeros in data files?

Prepare data for statistical analysis?

Use a codebook?

Use alphanumeric codes?

Use blank spaces in data files?

A "case"?

A codebook?

A database?

A numeric code?

A self-coding variable?

A variable?

An open-ended or non-structured response?

Coding for maximum information?

Coding?

Data entry?

Fixed-column format?

Free format?

Measurement?

Missing data?

Missing values?

Mutually exclusive categories (codes)?

The basic rules I should follow when coding variables?

The rules for naming variables?

Level: Beginning

Focus: Instructional (Detailed)

How Do I Prepare Data for Statistical Analysis?

This chapter concerns the preparation of data for analysis by a statistical program package such as SPSS® for Windows (Statistical Package for the Social Sciences), SAS® (Statistical Analysis System), or Number Cruncher® (NCSS). There are two components of this process. The first concerns the "coding" of data in a manner that permits analysis by statistical software, and the second concerns the process of "data entry." Our main focus is with the first of these two concerns. If the data are not properly prepared, the method used to enter the data into a computer file (i.e., a database) will not fix these problems. There are several potential data entry alternatives, as data may be entered into database management programs or spreadsheet programs or directly into statistical analysis programs through data entry options available with these programs. Data entry also can be achieved with a word processing program and then "read" into some statistical software programs; however, this is not usually the method of choice. If you are comfortable with a spreadsheet or database

An earlier version of this chapter was written by Carlene Nelson and Rae R. Newton. The authors wish to acknowledge Ms. Nelson's contribution to this chapter.

management program, we suggest that you use one of these to enter data. Remember that not all statistical programs can access data files created with a wide variety of software, so it is a good idea to investigate the data management capabilities of your statistical software before you enter your data.

In sum, when organizing data that have been collected for computer analysis, the user needs an understanding of how a data analysis program "reads" this information and reflects it back to the researcher in a meaningful way. The purpose of this chapter is to provide a guide for organizing data in a manner that permits a statistical program to read and analyze it. We begin with the basics; in the following paragraphs, we discuss cases, variables, values, codes, and the essential rules of data organization. Remember that most of what we say below has exceptions, and some would argue that there is a "better way." Our recommendations are designed to assist the novice and avoid the most common mistakes.

CASES: HOW DO I DEFINE A UNIT OF ANALYSIS?

Measurement is defined as the process of assigning numbers to characteristics of persons, places, events, or things that are observable. In the process of observing and measuring, each person, place, event, or thing may be considered a primary unit of analysis and may be measured on a variety of characteristics. We call these basic units **cases**.

In laboratory experiments studying the growth of malignant cells, each animal treated with a different therapy represents one case. In social science surveys, each respondent may be considered one case. In the government's examination of the repayment of student loans, each loan becomes one case. We can see, therefore, that in each research project there will be a specific number of cases that will be analyzed. This is often referred to as the study's number of cases, sample size, or simply "N."

HOW DO I USE VARIABLES AND VALUES TO DESCRIBE A CASE?

As we consider the measurement process, note that each object (i.e., each case) can be described with a variety of characteristics. Each of these characteristics is known as a **variable**. The variable itself may take on two or more different **values**. Values can be defined as categories that constitute the variable.

TABLE 1.1 How Variables Describe an Object	
Name of Variable	*Possible Values*
Shape	Round or oval
Color	White or brown
Construction material	Plastic or leather
Sport	Golf or football

For example, a ball (object) can be described in terms of its size (variable) as being "large" or "small" (values). In this instance, the variable SIZE is described rather crudely with the values large or small. SIZE also can be described utilizing more sophisticated values, such as centimeters to measure the diameter of the ball. Thus, the variable SIZE can describe the ball with either simple or exact categories. The different ways in which we choose to measure the size of the ball may reflect different **levels of measurement** and ultimately influence our choice of appropriate statistics.

Because the ball has more than one characteristic, it can be described with more than one variable. For example, in addition to its size, a ball can have the following characteristics or variables: shape, color, construction material, and sport in which it is used (see Table 1.1). Thus, each case may be described using multiple variables, and each variable is likely to contain many values. This hierarchy is illustrated in Figure 1.1.

WHAT ARE THE RULES I SHOULD
USE FOR NAMING VARIABLES?

Statistical software programs keep track of variables by using variable names. Typically, variable names are short mnemonics or numbers that refer to the variable. For example, the variable name for "age at first marriage" might be AGEWED. In this chapter, we follow the rules set forth for naming variables in the Statistical Package for the Social Sciences (SPSS). Within SPSS, a variable name (a) must start with a letter, (b) must be no longer than eight characters, and (c) may not contain any special characters such as blanks or commas. Thus, AGE WED would not be valid because it contains a space; VAR1 would be valid, as would V1, but VARIABLE1 would not be valid because it is too long (i.e., more

A Case (The Basic Unit of Analysis)

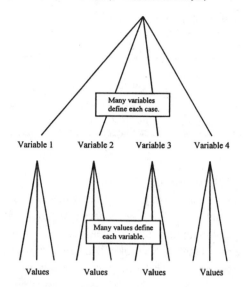

Figure 1.1. Illustration of Cases, Variables, and Values

than eight characters). Some researchers prefer to identify variables strictly by numbers such as VAR001, VAR002, and so on. Others prefer to use mnemonics such as SEX, AGE, INCOME, and ETHNIC. Note that most statistical software programs also allow the user to attach a longer "Variable Label" to the restricted variable name. Thus, the apparent confusion created by a long list of numbers (V1, V2, V3, etc.) can be alleviated by adding a complete list of variable labels.

WHAT IS DATA CODING?

Note that in the foregoing example of the variables used to describe a ball, the majority of the values of the variables would be words like "plastic" and "rubber" rather than numbers. Measurement, however, was defined as the process of assigning *numbers* to characteristics of objects we observe. In preparing data for the computer, we assign numbers to the categories of variables that do not normally take numerical values. We call the process of assigning numbers to each variable's values **the coding process**.

To illustrate, throughout this chapter we will use questions from a hypothetical questionnaire.

1. In the boxes below, write in your age at your last birthday.

☐☐

    The response to the question regarding respondent's age will be a number, not a word. Thus, the responses form the categories, with the youngest respondent's age forming the lowest value and the oldest respondent's age forming the highest value. A variable, such as AGE, is known as **self-coding** because the responses themselves form the numerical codes.

    Initially, the questionnaire is completed by either the respondent or the interviewer. In some cases, the second step may be to "code" the questionnaire by writing the code for each response in spaces provided for this purpose on the questionnaire. Finally, the data are entered into the computer, either directly from the questionnaire, or using the column of coded variables. This is explained more completely in the following pages. For example, if the respondent was 36 years old, he or she would write 36 on the questionnaire. This self-coding numerical response would then be entered directly into a computer file, or entered into a column on the right of the page and subsequently entered into a computer file.

    Another question from our hypothetical questionnaire asks about the respondent's sex.

2. What is your sex? (circle one)

    Male 0
    Female 1

    Note that in this question the response (value) is a word (male or female), not a number as in question 1, which asked respondent's age. To code this variable, we have arbitrarily assigned "0" to male and "1" to female and included these codes to the right of each response. (We could just have easily assigned "0" for female and "1" for male. The choice is up to the researcher within the limits of widely used conventions to be explained in a later section. We are following a practice known as "dummy coding" of dichotomous variables discussed in Chapter 8.)

    It is important to understand that the basic structure of the data we describe above can be reproduced by a variety of different software options, including spreadsheets, database managers, and word processors. In the spreadsheet format, the number of digits in the variable would be largely irrelevant, because each variable would be entered into a "cell" in the spreadsheet. Table 1.2

TABLE 1.2 Using Rows and Columns to Code Data

	Columns: Used to Code Variables		
		A	B	C
Rows: Used to Code Cases	1	32	1	215
	2	45	0	113
	3	33	1	165

TABLE 1.3 Using Rows and Columns to Code Data

	Columns: Used to Code Variables			
		A	B	C	D
Rows: Used to Code Cases	1	IDNUM	AGE	SEX	WEIGHT
	2	1	32	1	215
	3	2	45	0	113
	4	3	33	1	165

illustrates what this arrangement might look like in a spreadsheet program containing three variables that are coded into three columns, A, B, and C.

Note that the rows (1, 2, 3, etc.) and columns (A, B, C, etc.) of a spreadsheet do not necessarily represent variable names or case identification numbers. We recommend including these within spreadsheet files. Thus, if we rework the material above to include these suggestions, our spreadsheet file might look like Table 1.3.

HOW DO I CODE OPEN-ENDED RESPONSE CHOICES?

Sometimes a questionnaire item does not provide categories for all possible responses. For example, examine question 3 from the hypothetical questionnaire:

TABLE 1.4 Coding Frequency Data		
Race	*Frequency*	*Code*
Mexican American	1111 1111 1111 1111 1111	3
Spanish	1111 1111 1	4
Native American	1111 111	5

3. What race do you consider yourself? (circle one)

White............................ 1
African American 2
Other (specify)

When this question about race was constructed, it was assumed that the majority of the responses would fall into either the "White" or "African American" category. If the respondent's race is other than "White" or "African American," such as "Native American," he or she would write the response in the space provided:

Other (specify)
Native American

This type of response is known as **open-ended** or **nonstructured**.

In coding the answers to an open-ended response category, the reader would complete the following steps.

Step 1. Record all the responses on a tally sheet to determine the names of races designated as "Other." It is also recommended that you record the number of times each race occurs (frequency). For example, a respondent may have indicated that he or she is "Hispanic (Mexican American)" or "Hispanic (Spanish)." If the total number of "Hispanics" is much greater than the remaining races listed in the "Other" category, you may decide to break the Hispanic category down further into "Mexican American" and "Spanish" instead of leaving both of them in the "Hispanic" category.

Step 2. Assign additional codes to the new categories. For example, the tally sheet may resemble Table 1.4.

Step 3. Finally, return to the questionnaire and write in the appropriate code for the "Other" category. For example, if a respondent had answered the question as "Native American" and you had coded "Native American" as "5," you would place a "5" in the right-hand column.

3. What race do you consider yourself? (circle one)

    White............................ 1
    African American 2
    Other (specify)

    Native American   Race__5__

HOW DO I CODE
OPEN-ENDED QUESTIONS?

Open-ended questions are not as simple to code as the above illustration of an open-ended response. Consider questions 4 and 5 from the hypothetical questionnaire, in which people were asked their preference for burial or cremation as a method of body disposition, and then asked the reason for their choice.

4. If you were to die within the next few years, do you feel positive, neutral, or negative about:

    A. Burial
        Positive 1
        Neutral 2
        Negative 3

    B. Cremation
        Positive 1
        Neutral 2
        Negative 3

5. A. Why burial? (within the next few years)   Whybur _____
   B. Why cremation? (within the next few years) Whycrem_____

The answers to question 5 may be longer and more complicated than the open-ended response regarding race. For example, to the question "Why burial?", the following are several possible answers:

Burial gives me something to grieve over.

Burial is what has been practiced in my family.

My Catholic training in childhood taught me burial is the method of body disposition that most Catholics practice.

The idea of burning frightens me.

Because the answers are more complex, each response needs to be examined (i.e., **content analyzed**) to determine the underlying motivation. (A well-written introduction to content analysis is contained in Bruce L. Berg's [1998] book *Qualitative Research Methods for the Social Sciences*.) In other words, the researcher asks "Why did they say what they did?" These motivations then become the categories into which the responses fall. Let's consider the four responses to the question "Why burial?"

"Burial gives me something to grieve over." This answer might reflect the respondent's need for burial to aid him or her in resolving grief. Therefore, "aid in grief resolution" may be an appropriate category for this response.

"Burial is what has been practiced in my family." It appears that burial has been a part of this person's family tradition. Therefore, "family tradition" may be an appropriate category for this response.

"My Catholic training taught me burial is the method of body disposition that most Catholics practice." This person's answer appears to suggest the influence of early religious training. Therefore, "religion" may be an appropriate category for this response.

"The idea of burning frightens me." It appears that this person's motive for choosing burial is a fear of cremation. Thus, "fear of burning" may be an appropriate category for the response.

As you examine each case to determine the underlying motivation, you may discover that some motivations are common to more than one respondent. For instance, if there are 100 cases in the study, there will not be 100 response categories for the question "Why burial?" Instead, you may find that 10 people prefer burial for religious reasons, 20 people may be influenced by family tradition, 15 people fear burning, and so on. You may end up with several answers that are not at all alike and that do not seem to fit into any of the response categories created by the motivations. You may find it necessary to classify these types of responses as "miscellaneous motivations."

After you assign codes to the response categories that have been created, your tally sheet may resemble Table 1.5. Return to question 5 on each questionnaire

TABLE 1.5 Coding Frequency Data

Motivations for Burial	Frequency	Code
Aid in grief resolution	1111 1111 111	1
Family tradition	1111 1111 1111	2
Religion	1111 1111 1	3
Fear of burning	1111 111	4

and record the category into which the response falls, as well as the code for that category.

5. A. Why burial? (within the next few years)

"Burial gives me something to grieve over."
(respondent's answer)

Aid in grief resolution ___   Whybur___1___
(coder's notes)

5. A. Why burial? (within the next few years)

"Burial is what has been practiced in my family."
(respondent's answer)

Family tradition ___   Whybur___2___
(coder's notes)

5. A. Why burial? (within the next few years)

"My Catholic training taught me burial is the
method of body disposition that most Catholics practice."
(respondent's answer)

Religion ___   Whybur___3___
(coder's notes)

5. A. Why burial? (within the next few years)

"The idea of burning frightens me."
(respondent's answer)

Fear of burning ___   Whybur___4___
(coder's notes)

WHAT ARE THE BASIC RULES OF DATA ORGANIZATION?

Data can be organized in a number of ways into a format that can be understood by a computer program that calculates statistics. An individual with previous knowledge of this subject may have his or her own particular method for preparing and inputting data. For a beginner, we suggest a strategy that has been proven, through experience, to maximize efficiency and minimize errors.

The following is a list of "basic rules of data organization." These rules are important. The longer you work with data entry, the more you will appreciate their value.

Rule 1. All data must be numeric!

In the questionnaire, question 2 asks:

2. What is your sex? (circle one)

    Male 0
    Female 1

It may appear just as easy to code the values for sex as "M" and "F" as it is to code them "0" and "1." Such a practice, however, may prove troublesome when defining the data file for a statistical program. It is much easier to change variables that do not have numeric response choices, such as sex, to numerically coded variables before inputting the data rather than afterward. We strongly recommend this as a standard procedure.

Rule 2. Each variable must occupy the same location for each case.

This means each item of information is entered in exactly the same position (column) or cell location in a spreadsheet for each case in a row of data. When using data that is output from a spreadsheet, or data entered in a word processor and output as an ASCII (i.e., simple text) file, some statistical programs require that data be "delimited." This means that a comma or blank space (i.e., a delimiter) be placed between each pair of variables. More versatile programs can "read" or access data from files that use a **fixed-column format** and are not delimited. We strongly recommend that even if delimiters are used in a data file, the file be constructed in fixed-column format. This is the essence of Rule 2.

For example, look at the first and last items of information in the sample questionnaire starting on page 24.

Respondent's ID No. ___ ___ ___   Id ___ ___ ___
                                                      1   2   3

5e. (Satisfaction with) your health and physical condition

    A great deal 1
    None 7

Note that the respondent's ID number occupies the first three columns (positions) of every case, whereas the last variable (5e) occupies column 18 for each case.

The fixed-column format is the most commonly used format for statistical data because it makes sense that when using a spreadsheet, a specific column of cells be allocated to one variable; however, another available format is the **free format**. In this format, you do not need to place the data in the same location for each case. It is necessary only to enter the data in the correct sequence, with one or more delimiters between each data entry.

In the above example, if you used a free format, the respondent's ID number would be entered as the first variable in the sequence and the variable 5e would be entered last in the sequence. The free format is helpful in that you need not pay attention to column position when entering data. The disadvantage of the free format is that it is easy to omit data and extremely difficult to correct mistakes. We therefore recommend against both the use of word processors as a data entry option and the use of free format data.

Rule 3. All codes for a variable must be mutually exclusive.

By "mutually exclusive," we mean that each case can be classified into one and only one category (value) on a particular variable. For example, question 6 in the hypothetical questionnaire asks about the respondent's marital status.

6. Are you currently married, widowed, divorced, separated, or have you never been married? (circle one)

    Currently married 1
    Widowed 2
    Divorced 3

Separated 4
Never married 5

Let's suppose we change the wording of the above question to read:

6. Are you currently married, not married, or do you have children?

Currently married 1
Not married 2
Have children 3

It is possible for a respondent who has children to fit into both the "married" and "have children" categories. Because the respondent can fit into more than one category, the categories are not mutually exclusive. To satisfy the rule for mutually exclusive codes, it would be necessary to eliminate the category "have children" and add additional new categories. The real issue here is that two variables are being assessed with one question (i.e., number of children and marital status). It is a good idea to give your data collection strategy a thorough pretest to make sure you are aware of all the variables that can be derived from your instruments.

Rule 4. Each variable should be coded to obtain maximum information.

Let's consider the question regarding the respondent's marital status referred to in the foregoing paragraph. When the values for respondent's marital status are "married" and "not married," people can be classified into two broad categories. This provides us with only a minimum amount of information. All we know is that the respondent is either married or single. A more precise definition of current marital status would include the values

Currently married   = 1
Widowed   = 2
Divorced   = 3
Separated   = 4
Never married   = 5

By providing more precise categories, a maximum amount of information is obtained for the marital status variable.

In another example of coding a variable to obtain maximum information, consider the question related to respondent's age.

1. In the boxes below, write in your age at your last birthday.

Let's suppose we change the wording of the above question to read:

1. How old were you at your last birthday?

Younger than 18........... 1
18-30............................ 2
31-65............................ 3
66 or over.................... 4

When age is grouped into three categories, the exact age of the respondent is no longer available. Although the three-category classification may be helpful for identifying stages of adult development, it is not useful if the exact age of the subject is required. In addition, different theorists may have distinct views regarding what the stages of adult development are. If the data are entered to represent the respondent's exact age, the computer can be given the task of creating an endless variety of age groupings. Once the data have been grouped and entered as above, however, the computer cannot determine the exact age of the respondent or restructure the data. Thus, it is possible to "collapse" data in numerous ways, but not possible to "expand" data that have been collapsed prior to data entry.

Rule 5. For each case, there must be a numeric code for every variable.

Throughout the coding process, it is necessary for a code to be assigned to every variable. In some cases, there will be missing information. For example, there may be no information for marital status for a particular case, but codes are assigned even if the data are not available. Missing information will be explained in greater detail in the following section. At this point, suffice it to say that if missing information was not coded and the cell location in a spreadsheet were left blank, there could be confusion later in the analysis process. For example, if there was not a code for a particular variable, it might be difficult to determine whether the information was indeed missing or whether a valid code inadvertently had been left out.

When one enters data into a spreadsheet program or directly into a statistical program, it may be difficult to retrieve the coding system without some directory to the meaning of the various codes. As an example, consider the variable "Ethnicity." Initially, a researcher may have decided to code Whites as 1, Hispanics as 2, Asians as 3, and African Americans as 4. Without labels linking each code with its respective category, however, it becomes impossible to know that Hispanics were originally coded as "2." Such information is contained in a document known as a **codebook**. A codebook for the sample questionnaire can be found at the end of this chapter. We will discuss codebooks in greater detail in the section describing this codebook.

HOW DO I DEFINE AND CODE MISSING DATA?

As we gather data for each case in our study, some items of information, for one reason or another, often are not available. This results in an incomplete set of data for some cases.

The following is a list of possible explanations for missing data.

The person refused to answer.

The person is not home.

During a phone interview, the person hangs up.

The interviewer skips three or four questions.

An experimental animal dies halfway through the study.

A recording instrument fails.

Similarly, data collected by other persons—the U.S. Census, for example—may have certain pieces of information not available for every case. Missing data must be included in the coding system. If the data are excluded, we have broken Rule 5: "For every case, there must be a numeric code for every variable." In the SPSS statistical program, for example, any code can be designated for a missing value. A generally accepted convention is to use "9" for variables whose maximum value is no longer than 1 digit or greater than 8, called a 1-column variable; "99" for those variables whose values are greater than 8 but less than 98 (2-column variables); and so on for 3-digit (999) and larger variables.

For example, question 7 from the hypothetical questionnaire represents a 1-column variable (there are less than 9 response categories, excluding missing values).

7. How far did you go in school? (circle one)

Less than eight years......................... 1
Elementary school graduate.............. 2
Some high school 3
High school graduate......................... 4
Some college or AA degree.............. 5
BA, BS, RN (college degree) 6
Some graduate school...................... 7
Completion of advanced degree 8

Because there are less than nine response categories, we can designate "9" as the code for missing data. It is particularly important to point out here that assigning missing values during the data entry phase of a research project must be followed by informing the statistical software of these missing value codes. The software won't automatically "know" that a 9 or 99 represents a missing value. During a "data definition" session, we inform the statistical software of our variable names and missing value codes.

HOW DO I CODE NONSPECIFIC RESPONSES?

Sometimes a researcher believes it is necessary to analyze nonspecific responses instead of assigning them as missing data. For example, look at question 8 in the hypothetical questionnaire.

8. Do you believe there is a life after death? (circle one)

Yes 1
No 2
Undecided.................... 8

"I don't know," "I'm not sure," and "Undecided" all represent nonspecific responses. In the SPSS system, just as any code can be used to designate missing data, so too can any code be used to designate nonspecific responses. A generally accepted convention for nonspecific responses is to use "8" for 1-column variables, "98" for 2-column variables, and so on.

As we mentioned earlier, coding the variable is up to the discretion of the researcher (within the specifications of the statistical program being used). A case in point is question 9 from the hypothetical questionnaire.

9. How often do you attend religious services? (circle one)

Never 0
Less than once a year.. 1
About once a year........ 2
Several times a year 3
About once a month..... 4
2-3 times a month 5
Nearly every week 6
Every week 7
Several times a week... 8
No response 9

Note that "8" is a code representing the category "several times a week." It was not reserved for a nonspecific response such as "Undecided." If the researcher wished to include an undecided code, creating 11 response categories (including the nonspecific response and missing value), the variable would become a 2-column variable and the codes would become the following:

Never   = 00
Less than once a year = 01
About once a year   = 02
Several times a year   = 03
About once a month   = 04
2-3 times a month   = 05
Nearly every week   = 06
Every week   = 07
Several times a week   = 08
Undecided   = 98
No response   = 99

Certain coding rules are not as universally followed as others. Where there is a choice involved, rules regarding missing responses are more generally followed than for nonspecific responses.

HOW DO I DEAL WITH
BLANKS AND ZEROS?

Blanks generally are not used to designate missing data. (Note that Rule 5 of data organization would be violated; that is, "For each case, there must be a numeric code for every variable.")

Zero codes also are not typically used to indicate missing data, because zeros frequently are used to represent "zero" of something. For example, in question 9, "How often do you attend religious services?", zero (0) represents "never."

We recommend that (a) blanks *never* be used in a data file, except as delimiters, and (b) zeros be used as missing value codes only when common sense dictates that a zero would be more appropriate than a 9.

This concludes our discussion of organizing data for computer input. In the following section, we will present a detailed example using a new questionnaire, a codebook, and the basic rules of data organization.

A DETAILED EXAMPLE OF DATA
ORGANIZATION AND CODING

In this section, we want to check your understanding of the basic rules of data organization, using a questionnaire to construct a codebook. Recall that a codebook is an index to your coding system. In it, you detail the names of your variables, a description of each variable (a short reference or a detailed description if necessary), and the codes and code categories you have devised, including the codes for missing and nonspecific responses. In addition, depending on the form of the data, you might include the column location of each variable. Codebooks vary widely in structure and content, but most major research organizations, such as the National Opinion Research Center, the Gallup and Roper organizations, the U.S. Census, and the Interuniversity Consortium for Political and Social Research distribute their data accompanied with codebooks. Without these, the structure of the data would remain a mystery. Sometimes codebooks include the complete text of every question and a univariate data distribution; in other cases, they are more simple. The following pages present a five-item questionnaire. Alongside each page of the questionnaire is a blank codebook.

Try to construct the codebook based on the information presented in the questionnaire. For assistance in coding, return to the section "What Are the Basic Rules of Data Organization?", where data organization for computer input is discussed. The following is a list of helpful guidelines:

1. *Variable Description:* This is a brief description of the variable. It should enable you to readily identify the questionnaire item. For example, "pornography laws" may be an adequate variable description to identify questionnaire item 1.

2. *Variable Name:* We will adopt the specifications of the Statistical Package for the Social Sciences (SPSS) and limit variable names to eight characters or less, with no blank spaces between the characters. For example:
   a. The variable name for respondent's ID number can be "IDNUM" but not "ID NUM."
   b. Respondent's age can be "AGE" but not "RESPONDENT'S AGE."
   c. SPSS also allows alphanumeric variables names; however, the first character of the variable name must be an alphabetic letter. For instance, if we have 10 variables, the list of variables could be V01, V02, . . . V10 or VAR01, VAR02, . . . VAR10.

3. *Value Labels:* These are merely the names of the response categories and their respective codes. You will note that all response categories for the remaining questions have been coded, with the exception of respondent's ID number and question 2, which are both self-coding.

4. *Cell/Column Number:* With respect to the cell or column number, note whether the data for an item to be input into the computer occupies one, two, or three positions in the row (i.e., is this a one-, two- or three-column variable?). For example, the first questionnaire item, respondent's ID number, occupies three spaces or columns in the row (1, 2, and 3). If you are using a spreadsheet program, the ID number would simply occupy cell A, and the ID number for the first case would be in cell 2A. These letter/number combinations form "cells" that define the location of a variable of any size, for any case. In this situation, the "column number" would be replaced by the column letter, and each column would represent a complete variable.

QUESTIONNAIRE CODEBOOK

Respondent ID No. (3 columns)	Question Number	Variable Description	Variable Name	Value Label	Cell/ Column Number

1. Which of these statements comes closest to your feelings about pornography laws? (circle one)

   There should be laws against the distribution of pornography whatever the age 1

   There should be laws against the distribution of pornography to persons under 18............................ 2

   There should be no laws forbidding the distribution of pornography 3

2. On the average day, about how many hours do you personally watch television? Enter number of hours _____

3. Here are four statements regarding the role of the working mother. Please circle whether you strongly agree, agree, disagree, or strongly disagree with each statement.

   A. A working mother can establish just as warm and secure a relationship with her children as a mother who does not work.
   Strongly agree 1
   Agree 2
   Disagree 3
   Strongly disagree...... 4

   B. It is more important for a wife to help her husband's career than to have one herself.
   Strongly agree 1
   Agree 2
   Disagree 3
   Strongly disagree...... 4

   C. A preschool child is likely to suffer if his or her mother works.
   Strongly agree 1
   Agree 2
   Disagree 3
   Strongly disagree...... 4

QUESTIONNAIRE CODEBOOK

Question Number	Variable Description	Variable Name	Value Label	Cell/ Column Number

D. It is much better for everyone involved if the man is the achiever outside the home and the woman takes care of the home and family.
Strongly agree 1
Agree 2
Disagree 3
Strongly disagree...... 4

4. In 1996, you remember that Clinton ran for President on the Democratic ticket against Dole for the Republicans and Perot as an Independent. Do you remember for sure whether or not you voted in that election?

Voted......................... 1
Did not vote 2
Ineligible.................... 3

A. IF VOTED: Did you vote for Clinton, Dole, or Perot?
Clinton....................... 1
Dole 2
Perot 3
Other candidate
(Specify:_____) ___

B. IF DID NOT VOTE OR INELIGIBLE:
Who would you have voted for, for President, if you had voted?
Clinton 1
Dole 2
Perot 3
Other candidate
(Specify:_____) ___

5. For each area of life circle the number that shows how much satisfaction you get from that area.

A. The city or place you live in.
A very great deal.. 1
A great deal.......... 2
Quite a bit............. 3
A fair amount........ 4

QUESTIONNAIRE CODEBOOK

Question Number	Variable Description	Variable Name	Value Label	Cell/ Column Number

Some 5
A little 6
None 7

B. Your non-working activities—hobbies and so on.
A very great deal 1
A great deal 2
Quite a bit 3
A fair amount 4
Some 5
A little 6
None 7

C. Your family life.
A very great deal 1
A great deal 2
Quite a bit 3
A fair amount 4
Some 5
A little 6
None 7

D. Your friendships.
A very great deal 1
A great deal 2
Quite a bit 3
A fair amount 4
Some 5
A little 6
None 7

E. Your health and physical condition.
A very great deal 1
A great deal 2
Quite a bit 3
A fair amount 4
Some 5
A little 6
None 7

Congratulations! Your codebook should be completed, and now can be used as a "dictionary" to describe the questionnaire. In the following pages, we present our version of the completed codebook. The following numbered descriptions correspond to the superscripted numbers in the codebook:

1. We chose to use alphanumeric variable names in this codebook (V01). Variable names that suggest the name of the variable itself may also be used. For example, we could have used IDNUM as the variable name for respondent's ID number.

2. Note that all variables have been assigned a specific cell (spreadsheet) and column (ASCII text) location in the line of data. This will hold true for each case and, therefore, fulfills Rule 2 of data organization.

3. The value labels are a description of the response categories and their respective codes. Some statistical packages may limit the length of labels.

4. We reserve "8" to designate nonspecific responses and "9" for missing data. See Rule 5 of data organization.

5. The codes are all numeric and thus satisfy Rule 1 of data organization.

CODEBOOK

Question Number	Variable Description	Variable Name	Value Label	Cell/Column Number
ID No.		V01[1]	Self-coding 999 Missing value	(A) 1-3[2]
1	Porno laws	V02	1 Porno laws for all ages[3] 2 Porno laws for under 18 3 No porno laws 8 Don't know 9 Missing value[4]	(B) 4
2	Hours spent watching TV	V03	Self-coding 98 Don't know 99 Missing value	(C) 5
3A	Working mother can have warm relationship with children	V04	1 Strongly agree 2 Agree 3 Disagree 4 Strongly disagree 8 Don't know 9 Missing value	(D) 7
3B	Wife helps husband's career rather than her own	V05	1 Strongly agree 2 Agree 3 Disagree 4 Strongly disagree 8 Don't know 9 Missing value	(E) 8
3C	Preschool child suffers if mother works	V06	1 Strongly agree 2 Agree 3 Disagree 4 Strongly disagree 8 Don't know 9 Missing value	(F) 9
3D	Man works—woman cares for home and family	V07	1 Strongly agree 2 Agree 3 Disagree 4 Strongly disagree 8 Don't know 9 Missing value	(G)10
4	1996 Presidential election	V08	1 Voted 2 Did not vote 3 Ineligible 8 Don't know 9 Missing value	(H) 11

Question Number	Variable Description	Variable Name	Value Label	Cell/Column Number
4A	If voted	V09	1 Clinton 2 Dole 3 Perot 4 Other (specify) 8 Don't know 9 Missing value	(I) 12
4B	If did not vote or ineligible, for whom would you have voted	V10	1 Clinton 2 Bush 3 Perot 4 Other (specify) 8 Don't know 9 Missing value	(J) 13
5A	Satisfaction from place you live	V11	1 A very great deal 2 A great deal 3 Quite a bit 4 A fair amount 5 Some 6 A little 7 None 8 Don't know 9 Missing value	(K) 14
5B	Satisfaction from hobbies	V12	1 A very great deal 2 A great deal 3 Quite a bit 4 A fair amount 5 Some 6 A little 7 None 8 Don't know 9 Missing value	(L) 15
5C	Satisfaction from family life	V13	1 A very great deal 2 A great deal 3 Quite a bit 4 A fair amount 5 Some 6 A little 7 None 8 Don't know 9 Missing value	(M) 16

Question Number	Variable Description	Variable Name	Value Label	Cell/Column Number
5D	Satisfaction from friendships	V14	1 A very great deal 2 A great deal 3 Quite a bit 4 A fair amount 5 Some 6 A little 7 None 8 Don't know 9 Missing value	(N) 17
5E	Satisfaction from health	V15	1 A very great deal 2 A great deal 3 Quite a bit 4 A fair amount 5 Some 6 A little 7 None 8 Don't know 9 Missing value[5]	(O) 18

How Do I Examine Data Prior to Analysis?

How do I . . . ? When should or shouldn't I . . . ?

Define the shape of a distribution?	Examine data prior to analysis?
Discover coding and data entry errors?	Use data with visual aids?
Display data visually?	Use a crosstabulations?
Examine a distribution?	Use a scatterplot?
Examine data from two variables at once?	Trace errors in data entry?

What is . . . ? What are . . . ?

A bivariate table?	A unimodal distribution?
A box and whisker plot (a boxplot)?	A z score?
A contingency table?	An outlier?
	Data screening?
A crosstabulation?	Kurtosis?
A histogram?	Linear relationships?
A normal distribution?	Negative relationships?
A percentage difference?	Non-linear relationships?
A scatterplot?	Positive relationships?
A skewed distribution?	Skew?
A stem-and-leaf diagram?	Strength of relationship?
A symmetrical distribution?	

Level: Beginning

Focus: Instructional

CHAPTER **two**

How Do I Examine Data Prior to Analysis?

Imagine mounting a large data collection effort, entering all the data into a computer following the guidelines in the first chapter, and then facing the question, "What do I do now?" Some researchers might be tempted to jump immediately to the testing of hypotheses using multivariate techniques, whereas others often don't know where to start. This chapter addresses the question of where to start with data analysis. There are two goals of this chapter. First, the procedures we suggest, though relatively easy to follow and understand, serve as the second step in a quality control process that protects data analysts from disaster. (The first step occurs at the data entry point and was discussed in Chapter 1.) Second, the blueprint for analysis that we outline here is central to the plan of this entire book. The well-prepared data analyst should be familiar with the basics outlined here before proceeding further.

It is naive to believe that once the data are collected, all the hard work is done, because the computer will simply "crunch the numbers" and answer all our questions. In fact, just as with qualitative analysis, one needs to get a feel for one's data and become comfortable with them before jumping into full-scale

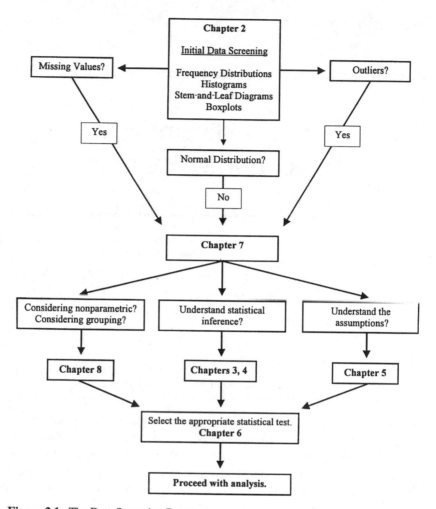

Figure 2.1. The Data Screening Process

hypothesis testing. This process starts from the ground up, first examining one variable at a time (univariate analysis) and building from there toward more complex statistical applications, if necessary and appropriate (multivariate analysis).

Examination of the characteristics of one's data is partly a visual activity, and we emphasize the visual inspection of data in this and other sections. A thorough inspection of data also makes use of tabular and mathematical manipulations. When we thoroughly understand the properties of our data, many of the questions regarding the ability of the data to meet the assumptions of sophisticated statistical analyses will be answered. In this chapter, we begin by suggesting strategies for examining the distribution of a single variable, particularly the ability of that variable to approximate a normal distribution. We follow this by suggesting methods for examining bivariate distributions. This process is likely to indicate issues or problems concerning the data that require some "adjustments" before moving on. These include methods for detecting and correcting for data containing missing observations and extreme scores or outliers. These topics are discussed in greater detail in Part III: Issues Related to Variables and Their Distributions. The flow chart in Figure 2.1 describes the data screening process and the chapters where specific issues are considered.

As suggested above, the initial step in the exploration of a data set is to examine the distribution of every variable that composes that data set. The starting point of this process is most typically the examination of the frequency distributions of the variables in question. This is necessary regardless of whether or not the variables are categorical or continuous and regardless of what statistical operations we ultimately conduct. We present frequency distributions throughout this book, and most computer programs provide distributions with a similar structure, that can be modified by the user. We will be using distributions as produced by hand, and by SPSS® for Windows®. For example, in Table 2.1 we present the distribution of "Age When First Married" from the National Opinion Research Center General Social Survey of 1995, as produced by SPSS® for Windows 95®. (Note that printouts from SPSS will differ slightly in their appearance from the reproductions in this book, but the data contained will be identical.) Note that the table has a column showing the valid codes (the ages) and the missing values codes (0 for NAP, and 99 for NA). The 286 NAP codes stand for the "not applicable" cases, those who were never married. The table contains columns of frequencies, percentages, valid percentages, and cumulative percentages. The difference between the percentages and valid percentages results from the missing values; the percentage column includes them, but the valid percentage column does not.

There are two sorts of things to look for during a first glance at a frequency distribution such as that in Table 2.1. First, we should attempt to get a preliminary sense of the shape of the distribution. Note that it appears that this distribution has a few cases who were married after 40, up to an age of 58. This is referred to as "positive skew" because the upper tail of the distribution tends to be pulled outward by a few extreme cases. The term **skew** as used to describe a distribution

TABLE 2.1 Distribution of Age When First Married

Age When First Married			
		Valid	Cumulative
	Frequency	Percent	Percent	Percent
Valid 13	1	.1	.1	.1
14	4	.3	.3	.4
15	7	.5	.6	1.0
16	32	2.1	2.7	3.7
17	43	2.9	3.6	7.2
18	118	7.9	9.8	17.1
19	129	8.6	10.7	27.8
20	121	8.1	10.1	37.9
21	132	8.8	11.0	48.8
22	96	6.4	8.0	56.8
23	82	5.5	6.8	63.6
24	82	5.5	6.8	70.5
25	72	4.8	6.0	76.5
26	61	4.1	5.1	81.5
27	49	3.3	4.1	85.6
28	27	1.8	2.2	87.9
29	34	2.3	2.8	90.7
30	25	1.7	2.1	92.8
31	18	1.2	1.5	94.3
32	21	1.4	1.7	96.0
33	10	.7	.8	96.8
34	5	.3	.4	97.3
35	7	.5	.6	97.8
36	6	.4	.5	98.3
37	3	.2	.2	98.6
38	4	.3	.3	98.9
40	3	.2	.2	99.2
41	1	.1	.1	99.3
42	2	.1	.2	99.4
43	1	.1	.1	99.5
45	1	.1	.1	99.6
47	1	.1	.1	99.7
49	1	.1	.1	99.8
50	1	.1	.1	99.8
54	1	.1	.1	99.9
58	1	.1	.1	100.0
Total	1202	80.1	100.0	
Missing 0 NAP	286	19.1		
99 NA	12	.8		
Total	298	19.9		
Total	1500	100.0		

generally refers to the distribution's deviation from a perfectly symmetrical shape. It is easy to see from this distribution that most people were first married between the ages of 18 and 21, but a few were first married in their 40s and 50s, hence the positive skew.

The second thing that researchers should be aware of when first examining their data is the possibility that errors in coding and recording were made. This is part of what researchers call "data cleaning." Sometimes data just don't make sense. For example, the distribution of age at first marriage shows that the youngest person married at the age of 13. This is probably a valid case, but if the youngest age was reported as 5 years, we might become suspicious of a data entry error. A good policy is to routinely produce the frequency distribution of every variable, even the identification numbers of cases, looking for these obvious errors. (When an ID number occurs twice, it may indicate that some data have been inadvertently entered twice or a data entry error was made, a very common occurrence with larger data sets.) Not all data entry errors are as easy to detect as obviously suspect codes or repeated ID numbers. If a 19 instead of a 29 was entered for marriage age, it is almost a certainty that the error would never be detected. The best precautions are first to enter the data carefully, and second to return to the original data and trace obvious errors from there.

HOW DO I DISPLAY MY DATA VISUALLY?

An examination of the frequency distribution might be sufficient for some types of research; however, if your goal is ultimately to utilize the variable in procedures involving statistical inference, you probably will also want to examine visual representations of the data. Most statistical packages do this, and some provide options, such as superimposing a normal curve over the histogram of the distribution, to assist with the evaluation of the shape of the distribution, as is shown in Figure 2.2 for age at first marriage.

The histogram clearly shows that the right tail of the distribution exceeds that of the superimposed normal curve (positive skew) and that the distribution is more "peaked" than a normal distribution. The general "peakedness" of a distribution is called **kurtosis**. Very flat distributions are called **platykurtic**, very peaked distributions, such as this, are called **leptokurtic**, and distributions approximating a bell shaped normal curve are called **mesokurtic**. Thus, the histogram helps us to see that the distribution of age of first marriage is both positively skewed and leptokurtic.

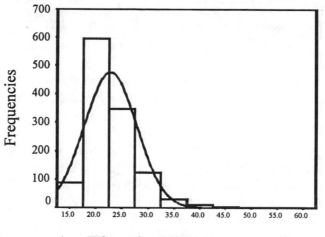

Age When First Married

Figure 2.2. Histogram of Age When First Married With Superimposed Normal Curve

A number of other choices exist for presentation of a univariate distribution of continuously distributed data. Two of these are the boxplot, or box and whisker plot, and the stem plot, or stem-and-leaf diagram. We illustrate each below, beginning with the boxplot for the age of marriage variable in Figure 2.3.

? What Is a Box and Whisker Plot?

A **box and whisker plot** contains a box whose boundaries represent the upper and lower quartiles of the distribution. This means that the middle 50% of the cases are contained within the box and the upper and lower 25% (below the first and above the third quartiles) are excluded. The horizontal line within the box represents the median, or the value of the case in the exact middle of the distribution (the point at the 50th percentile). The height of the box then can be interpreted as the distance between the first and third quartiles. The lines extending from the top and bottom of the box are called **whiskers** and represent the largest and smallest values that are not considered outliers. Outliers, represented with circles (O) and asterisks (*), are of two types, outliers (the circles) and extreme outliers (the asterisks). Outliers are values that range from 1.5 to 3 boxlengths from the edges; values more than three boxlengths from the edge are called extreme outliers. Different authors may define outliers differently, so be aware that this is only one recommended definition. Some computer programs

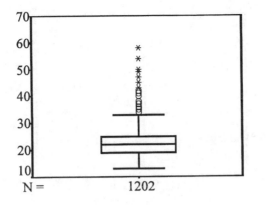

Figure 2.3. Boxplot for Age When First Married

also provide the identification numbers of the cases that are considered outliers. This should help the researcher return to the original data to examine the case, if necessary.[1] A more thorough discussion of how to deal with outliers can be found in Chapter 8. A diagram of this general framework for presenting box and whisker diagrams is presented in Figure 2.4.

Note that the boxplot for age at marriage (Figure 2.3) shows the median at approximately 22, with the upper (75th) and lower (25th) percentiles at 25 and 19, respectively. There is also a large number of outliers (38, to be exact). These are cases beyond a value of 34, which would be 1.5 boxlengths above the box boundary of 25. We obtained the value of 34 by taking a boxlength of 6 (25 − 19 = 6). Multiplying this value by 1.5 (6 × 1.5 = 9) and adding this to the value of the upper box boundary, we get 25 + 9 = 34. Values beyond 34 are classified as outliers, and values beyond 43, or 3 boxlengths (6 × 3 = 18; 25 + 18 = 43), are extreme outliers. Thus, the boxplot helps us visualize the distribution and clearly identifies the reason for the positive skew of this distribution, the large number of persons married beyond the age of 34.

A **stem-and-leaf diagram** is another way to visualize the distribution of a variable. Stem-and-leaf diagrams replace the bars of a histogram with values obtained from the data. A stem-and-leaf diagram is best explained with an example. We present the stem-and-leaf diagram of the age at marriage variable in Figure 2.5.

Note that each stem represents one of the age categories, from 13 to 33, and that each leaf, represented by a 0, represents 3 cases. The 38 outliers are not

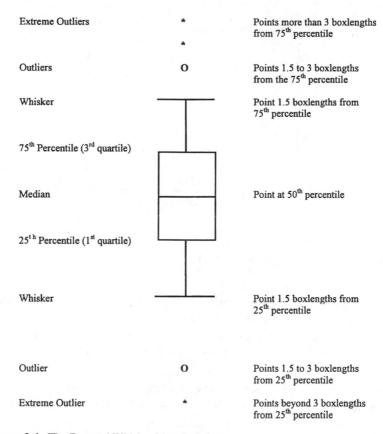

Extreme Outliers	*	Points more than 3 boxlengths from 75th percentile
	*	
Outliers	O	Points 1.5 to 3 boxlengths from the 75th percentile
Whisker		Point 1.5 boxlengths from 75th percentile
75th Percentile (3rd quartile)		
Median		Point at 50th percentile
25th Percentile (1st quartile)		
Whisker		Point 1.5 boxlengths from 25th percentile
Outlier	O	Points 1.5 to 3 boxlengths from 25th percentile
Extreme Outlier	*	Points beyond 3 boxlengths from 25th percentile

Figure 2.4. The Box and Whisker Plot (Boxplot)

included, and the one leaf in the 13-year-old-stem is represented by "&" because it contains less than 3 cases. Among other things, one can see that without the outliers, the distribution appears more symmetrical but still maintains its leptokurtic appearance. In Figure 2.6, we present a number of small graphs that depict the language that statisticians use to describe the shape and form of a distribution.

WHAT IS A NORMAL DISTRIBUTION?

The term "normally distributed" has appeared a number of times in the above discussion and will appear frequently throughout this book. A **normal distri-**

```
Frequency          Stem &   Leaf

     1.00       13 .  &
     4.00       14 .  0
     7.00       15 .  00
    32.00       16 .  00000000000
    43.00       17 .  00000000000000
   118.00       18 .  0000000000000000000000000000000000000000
   129.00       19 .  000000000000000000000000000000000000000000000
   121.00       20 .  00000000000000000000000000000000000000000
   132.00       21 .  000000000000000000000000000000000000000000000
    96.00       22 .  0000000000000000000000000000000000
    82.00       23 .  0000000000000000000000000000
    82.00       24 .  0000000000000000000000000000
    72.00       25 .  000000000000000000000000
    61.00       26 .  00000000000000000000
    49.00       27 .  0000000000000000
    27.00       28 .  000000000
    34.00       29 .  00000000000
    25.00       30 .  00000000
    18.00       31 .  000000
    21.00       32 .  0000000
    10.00       33 .  000
    38.00 Extremes    (>=34)

         Stem width:   1
         Each leaf:    3 case(s)

      &: denotes fractional leaves.
```

Figure 2.5. Stem-and-Leaf Diagram of Age When First Married

bution is a theoretical probability distribution and is a special case of a symmetrical, unimodal, bell shaped curve. This curve may be represented by real-world phenomena, but as shown in the examples above, many distributions deviate markedly from even a symmetrical, bell shaped structure. Virtually all statistics texts provide examples of normal curves and contain a table that defines the area underneath the curve. These areas are located through the use of z scores and directly correspond to probabilities. A z score is a "transformation" of a normal probability distribution in such a way that the mean of this distribution will be 0 and the standard deviation will be equal to 1. Because the z score transformation standardizes the distribution, the term "standard normal distribution" is used to describe these curves. Thus, a z score of +1 indicates the point on the horizontal axis that is one standard deviation above the mean, and a z score of -2 indicates a point two standard deviations below the mean. Figure 2.7 presents an example of a normal curve.

Note that the horizontal axis, labeled z, has positive and negative values around a mean of zero. These "z scores" represent standard deviations above and below the mean. In a normal distribution, the area between the mean and the z scores is known, and is defined by the "normal curve table." For example, in

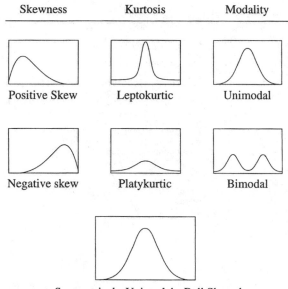

Figure 2.6. Distribution Shapes and Forms

Figure 2.7, the area between the mean and a z score of +1 is .3413 (34.13%). Because the curve is perfectly symmetrical, the area between the mean and a z score of –1 is also .3413. The area between a z score of +1 and –1 is thus the sum of these two areas, or .6826. Because the normal distribution is a probability distribution, the areas between any two points on the curve can be interpreted as probabilities. This point will become extremely important in our discussion of the logic of statistical inference that follows in the next section.

HOW DO I EXAMINE TWO VARIABLES AT THE SAME TIME?

Once we have a reasonable picture of the univariate distributions of our variables, we can begin to consider how these variables may work together. In other words, we can begin to think about relationships between variables. There are two basic, preliminary strategies for doing this that depend on the characteristics of the data. For two *categorical* variables with a small number of categories, either discrete or orderable discrete, we recommend a **crosstabula-**

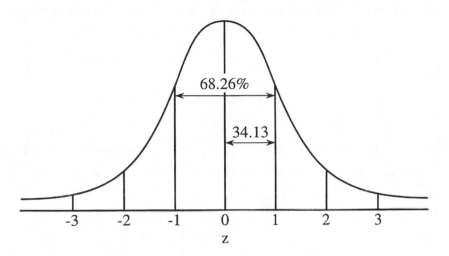

Figure 2.7. A Standard Normal Distribution

tion. For two *continuous* variables, the **scatterplot** is the method of choice. When one variable is continuous and one discrete, we may choose to compare boxplots or construct back-to-back stem-and-leaf plots. We briefly describe each of these below.

? What Is a Crosstabulation?

A **crosstabulation** (also called a bivariate table or contingency table) contains the joint distribution of two variables. The categories of each variable are laid out in the form of a square or rectangle containing rows (representing the categories of one variable) and columns (representing the categories of the second variable). Thus, if we wished to examine the bivariate relationship between sex and attitudes toward abortion, our table might appear as in Table 2.2.

The entries that occur in each cell will be the frequencies representing the number of times that each pair of values (one from each variable) occurs in the sample. These are called cell frequencies. The entries that occur at the end of each row or column are simply sums across the rows or columns and are called marginal frequencies. These represent the total number of times the category of the row variable or column variable occurred. By adding either the row marginal frequencies or the column marginal frequencies, you obtain the total N (total sample size). By widespread agreement, the conventional way to construct a table is with the independent variable at the top and the dependent variable at

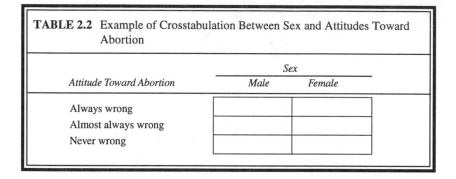

TABLE 2.2 Example of Crosstabulation Between Sex and Attitudes Toward Abortion

	Sex	
Attitude Toward Abortion	*Male*	*Female*
Always wrong		
Almost always wrong		
Never wrong		

the side of the table. Thus, if we believe that attitudes toward abortion are partly a function of sex (the only logical choice in this instance), then sex is placed at the top of the table and becomes the column variable.

Our goal is to examine the crosstabulation to get a sense of how the cases distribute themselves into the cells of the table (i.e., their joint distribution) and, second, to get a sense of whether or not the two variables are related. For example, examine the three parts of Table 2.3, all showing the distribution of cases into the sex by abortion crosstabulation.

The first part of the table illustrates a situation in which the examination of the relationship between the two variables would probably be unreasonable because, for whatever reason, the sample was composed mostly of women. With only 10 men in the sample, it probably would be unreasonable to conclude that sex and attitudes toward abortion are related or to calculate measures of statistical association. The second part of the table shows a case in which the cases do distribute among the cells and no cell frequency is exceedingly small, but the variables are unrelated. The third part of the table shows a case in which the cases distribute among the cells and the variables are related. How could we tell that the variables were unrelated in the second part of the table, but related in the third? The answer lies in how we standardized the tables by calculating the percentages. To calculate percentages, we may work within columns, within rows, or with the total N. Thus, the tables may be percentaged in three different ways. In fact, only one of these ways provides meaningful answers to the question of the relationship between the two variables. This is to calculate the percentages within the columns. When working within columns, the column total becomes the base number used to calculate the column percentage. For example, to find the percentage of male answers among those believing that abortion is always wrong, take the total number of male responses in that column

TABLE 2.3 Sample Outcomes for Attitudes Toward Abortion, by Sex						
Attitude Toward Abortion	*M*	*F*	*M*	*F*	*M*	*F*
Always wrong	4	804	380	760	380	380
Almost always wrong	3	477	70	140	310	80
Never wrong	3	209	50	100	310	40
Total	10	1,490	500	1,000	1,000	500

and divide it into the number of males in that column who believe that abortion is always wrong, and then multiply the result by 100. After we do this for each column, the final step is to compare across the column distributions within any one row. When we do this across the top row for the three examples in Table 2.3, we get the results shown in Table 2.4. We find that the difference in percentages between males and females is 14% for the first example, zero for the second, and 38% for the third. This becomes the basis for our conclusion that the variables are unrelated in the second table and related in the third. Note that the 14% difference across the top row of the first table would reduce to 4% or increase to 24% by simply shifting one case into (or out of) the top left cell. This is our basis for rejecting the notion that meaningful conclusions can be made from this table.

The point is that we cannot meaningfully simply compare the numbers of cases in each cell because the numbers of males and females are different. We must first standardize the distributions by calculating the column percentages.

The above method of describing categorical data in two-way tables is only a precursor to more advanced techniques of contingency table analysis. We refer the interested reader to models for the analysis of two-way tables described by Velleman and Hoaglin (1981) and Behrens (1997).

? How Do I Use a Scatterplot to Examine Continuously Distributed Bivariate Data?

When working with continuously distributed data, the questions are the same as when working with categorical data. Is it reasonable to examine this relationship and, if so, are the variables related? If we are interested in examining the relationship between two continuously distributed variables, one method to obtain a preliminary understanding of this relationship is to plot the values of

TABLE 2.4 Percentage Outcomes for Sex, by Attitudes Toward Abortion

Attitude Toward Abortion	M	F	M	F	M	F
Always wrong	40	54	76	76	38	76
Almost always wrong	30	32	14	14	31	16
Never wrong	30	14	10	10	31	8
Total	100	100	100	100	100	100
	(10)	(1,490)	(500)	(1,000)	(1,000)	(500)

NOTE: Numbers in parentheses are cell sizes.

the variables on a graph, called a **scatterplot** or **scatter diagram**. Place one variable (the dependent or *Y* variable) on the vertical axis, and the other (the independent or *X* variable) on the horizontal axis. Place appropriate scales to match the distribution of each variable on the axes and then place a dot on the graph to represent the location of each point on the graph. The example in Figure 2.8 shows the relationship between two scales, one assessing the level of exposure to violence and the other assessing symptoms of post-traumatic stress disorder.

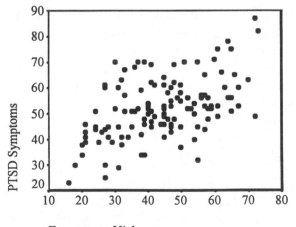

Exposure to Violence

Figure 2.8. Scatterplot of the Relationship Between Exposure to Violence and Symptoms of Post-Traumatic Stress Disorder

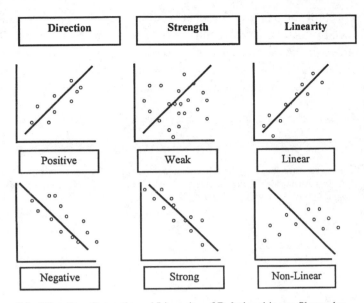

Figure 2.9. Direction, Strength, and Linearity of Relationships as Shown by Scatterplots

We can learn a number of things by inspection of the scatter diagram. First, we can determine if the relationship between the variable is positive or negative. **Positive relationships** represent relationships in which an increase in one variable is associated with an increase in a second. **Negative relationships** represent relationships in which an increase in one variable is associated with a decrease in a second. Second, we can determine if the relationship is linear or non-linear. **Linear relationships** are indicated when the pattern of dots on the scatter diagram appears to be straight, or if the points could be represented by drawing a straight line through them. Third, we can estimate the strength of the relationship between the variables. Strong relationships appear as those in which the dots are very close to a straight line; weaker relationships appear as those in which the dots are more scattered about a straight line, or farther away from that line. For the example in Figure 2.8, the relationship is both positive and linear, as the pattern of dots goes from lower left to upper right in a "cigar shaped" pattern that is typical of scatterplots representing linear positive relationships. The relationship also appears fairly strong. This judgment is supported by the value of the correlation coefficient of +.55. Figure 2.9 presents some examples of scatterplots with accompanying interpretations.

In conclusion, the first steps in data analysis begin with an examination of one's data, first one variable at a time, and then two variables at a time. These observations provide us with a picture of potential errors in data collection and entry, and the distributions of the variables. We then can explore the data through the use of bivariate analyses, using crosstabulations and scatterplots.

NOTE

1. Emerson and Strenio (1983) provide a detailed discussion of boxplots in exploratory data analysis. The interested reader is referred to this article and Hoaglin, Mosteller, and Tukey's (1983) *Understanding Robust and Exploratory Data Analysis*.

PART two

The Logic of Statistical Analysis

Issues Regarding the Nature of Statistics and Statistical Tests

In the first part of this book, we considered questions that relate to the general issue of preparing data for analysis and preliminary data screening. In Part II, we consider the logic of statistical testing. In Chapter 3, we begin with the general question of "What is statistical analysis?" We address statistics both as a field of inquiry and as a group of specific measures that one "calculates." We discuss the distinction between descriptive and inferential statistics and the logic underlying statistical inference. Finally, we discuss what it means to say that a finding is "statistically significant." This chapter is both an introduction and a "refresher" and should be read by anyone feeling incomplete or a bit out of date in his or her statistical training.

After years, even decades, of critique, "traditional" approaches to statistical inference are beginning to crumble under a barrage of new paradigms for analysis being proposed by a variety of social scientists. No book discussing statistical analysis would be complete without some discussion of both the logic of these critiques and an overview of the alternative approaches. We address this material in Chapter 4. Chapter 4 should be read by anyone not familiar with this material, even the reader fairly well versed in the logic of traditional statistical analysis discussed in Chapter 3.

In Chapter 5, we address the "assumptions" of statistical testing. This chapter is an extension of what was begun in Chapter 2, where we discussed how to take a first look at one's data. In this chapter, we expand further on the relationship between the characteristics of data and the assumptions underlying the use of statistical testing.

CHAPTER three

Traditional Approaches to Statistical Analysis and the Logic of Statistical Inference

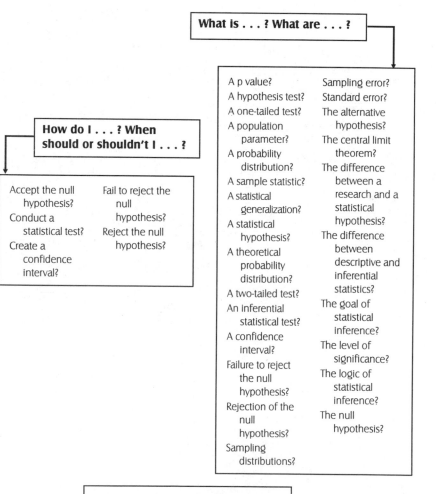

What is . . . ? What are . . . ?

A p value?
A hypothesis test?
A one-tailed test?
A population parameter?
A probability distribution?
A sample statistic?
A statistical generalization?
A statistical hypothesis?
A theoretical probability distribution?
A two-tailed test?
An inferential statistical test?
A confidence interval?
Failure to reject the null hypothesis?
Rejection of the null hypothesis?
Sampling distributions?

Sampling error?
Standard error?
The alternative hypothesis?
The central limit theorem?
The difference between a research and a statistical hypothesis?
The difference between descriptive and inferential statistics?
The goal of statistical inference?
The level of significance?
The logic of statistical inference?
The null hypothesis?

How do I . . . ? When should or shouldn't I . . . ?

Accept the null hypothesis?
Conduct a statistical test?
Create a confidence interval?

Fail to reject the null hypothesis?
Reject the null hypothesis?

Level: Beginning/Intermediate

Focus: Conceptual

three

Traditional Approaches to Statistical Analysis and the Logic of Statistical Inference

The American Heritage Dictionary (3rd ed., 1994) defines statistics as the mathematics of the collection, organization, and interpretation of numerical data, especially the analysis of population characteristics by inference from sampling. Clearly, the emphasis on mathematics and inference is critical, but for social and behavioral scientists, statistics is more likely to be viewed as a set of tools that assist with the organization, analysis, and interpretation of data. With the increasing availability of extremely powerful computers and ever simpler statistical program packages, researchers from all disciplines can perform sophisticated analyses with few thoughts to the intricacies and complexities of the mathematical manipulations that underlie them. The emphasis on statistics as "tools" rather than a branch of mathematical science shifts the focus to the task of answering empirical questions; however, with the broad diversity of questions scientists seek to answer and the wide variety of statistical techniques available comes the question of making an appropriate "match" of research question to statistical technique. As stated by Kachigan (1986), "The biggest challenge is to be able to recognize which types of analysis are most appropriate for a given situation, and how to interpret and apply the results" (p. 3).

In this chapter, we wish to accomplish three goals related to the above issue. First, we clarify the distinction between descriptive and inferential statistics. Second, we discuss what is commonly referred to as "the logic of inference." Third, we explore the meaning of the term "statistically significant." This chapter is a foundation for the remaining chapters, particularly those contained in Part II of this book. If you are unfamiliar with the distinctions mentioned above, you should reacquaint yourself with the material that follows.

WHAT IS THE DIFFERENCE BETWEEN DESCRIPTIVE AND INFERENTIAL STATISTICS?

Statistical analysis can be viewed as primarily concerned with two primary foci, the description of sample information, or **descriptive statistics**, and the generalization of sample information to a population, or **statistical inference**. The goal of descriptive statistics is to organize and describe data obtained from a sample of observations. This may involve the presentation of distributions in tables and graphs, data reduction techniques that summarize a large number of observations by reference to a single number, and the description of relationships between observations. Typically, data analyses include all the above types of descriptive techniques. For example, if you collected data from a sample that included the age of each respondent, you could present the distribution of age in a table or a graph, you could summarize the age data by reducing the age distribution to a single number representing the average age, and you could examine the relationship between age and other data you may have collected, such as education, income, or number of children. All the above methods of treating the data represent examination of the data by means of descriptive statistics.

Descriptive statistics that are used to describe samples—such as percentages, measures of central tendency (e.g., the mean), measures of variability (e.g., the standard deviation), and measures of strength of relationship (e.g., the correlation coefficient)—may be considered estimates of the value of that same measure in the population from which the sample was drawn. Thus, the sample mean is an estimate of the population mean, the sample standard deviation is an estimate

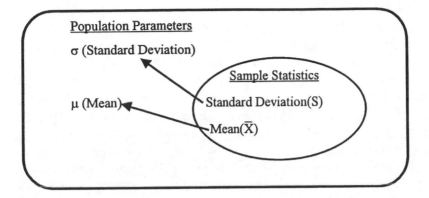

Figure 3.1. An Illustration of Inference From Sample to Population

of the population standard deviation, and the sample correlation is an estimate of the population correlation. This idea is illustrated by Figure 3.1.

Figure 3.1 illustrates a population from which a sample has been selected. The measures in Figure 3.1 that summarize characteristics of the population are referred to as **parameters** and are represented with Greek letters. The corresponding measures in Figure 3.1 that describe the sample are referred to as **sample statistics**, or simply statistics, and are represented with the common alphabet. Thus, Figure 3.1 shows the population mean (a parameter) represented with the Greek letter (mu, or μ) and the sample mean (a statistic) represented by $\overline{X}$ (bar X).

Assuming that the sample has been selected using an appropriate sampling design, one may use the information calculated from the sample to make inferences regarding the unknown characteristics of the population. The process of making such inferences is known as **statistical inference** and is based on probability theory (hence the heavy emphasis of probability and mathematics in some statistics courses). The next section considers the underlying logic of this process in more detail.

WHAT IS A STATISTICAL GENERALIZATION?

The fundamental idea that serves as the foundation of statistical inference is the notion of generalization. We wish to generalize from a measure calculated from a sample to the population from which that sample was drawn. To be convincing, such a generalization must have two features. First, it should be **accurate**, and second, it should be **precise**. By "accurate," we mean that we should have some level of confidence that the estimate is in fact correct. By "precise," we mean that the estimate is expressed within close and clearly specified limits. For example, a statement that exactly 56% of the American people support capital punishment is very precise but almost certainly incorrect, whereas a statement that between 10% and 90% of the American people support capital punishment probably is correct but is so imprecise that no useful information is provided. The goal of statistical inference is to make precise estimates of population parameters, with known risks of error based on observations from samples. To do this, we must develop a theory that relates sample statistics to population parameters. The foundation of this theory rests with our understanding of the nature of sampling and the errors that can occur during the sampling procedure.

? How Do Errors Arise in Sampling?

Identifying sources of error in samples begins with a consideration of the sampling process itself. Some errors occur because the nature of the sampling process is fatally flawed. Flawed sampling procedures are referred to as **biased**. Once the data are collected, there is little or nothing that can be done to correct for these fundamental mistakes in sampling design. Bias cannot be measured, and the selection of a larger sample does not correct for the fact that the sample is biased.

A second type of sampling error can be measured and controlled. This type of error occurs in samples that are selected randomly and is referred to as **random sampling error**. Random sampling error occurs because the process of sampling is in essence a process with limitations. These limitations occur because the entire population cannot be sampled. The smaller the sample, the more likely it is that random fluctuations in the selection of observations from the sample will result in deviations between the value of sample statistics and the population parameters they are selected to represent. Conversely, the larger the sample, the more likely it is that sample statistics will accurately estimate the population parameters. In fact, it can be shown that random sampling error is minimized when sample sizes increase and that the amount of this type of sampling error can be measured.

A **random sample** is defined as a sample drawn in such a manner that each and every object in the population has an equal chance of being selected. (More precisely, such samples are drawn so that all possible samples of the same size have an equally likely chance of being selected.) Through the selection of a random sample, one eliminates bias and permits the measurement of random sampling error. Random sampling error is measured through the use of a special distribution called a **sampling distribution**. We discuss sampling distributions in the next section.

WHAT ARE SAMPLING DISTRIBUTIONS?

When conducting research, scientists seldom take more than one sample from a population. This single sample becomes the basis upon which inferences are made. Consider for a moment the possibility of selecting numerous samples using identical random sampling procedures from the same population. We would now have multiple instances of whatever statistic we were interested in examining (i.e., a mean or a proportion or something else). The differences between these sample statistics might give us some notion concerning how well our sampling procedure was working. For example, if the sample statistics were very similar, we might gain some confidence in our sample estimates as representative of the population value, but if they were very different, we might question the reliability of these estimates. Now consider the possibility of repeating this procedure a larger number of times. Over time, we would be able to create a distribution of these sample statistics. Each sample of the same size would provide one observation (i.e., a statistic) to be included in this distribution. Such a distribution is called a **sampling distribution**, and assuming that we are using a continuously distributed variable, this distribution has a mean and a standard deviation. In more formal language:

A sampling distribution is the distribution of a sample statistic that would be obtained if all possible samples of the same size (N) were drawn from a given population.

When the statistic we are interested in is a mean, the sampling distribution would be the distribution of all means based on equal size samples drawn from the same population. The characteristics of the mean and standard deviation of this sampling distribution are described in a theorem known as the **Central Limit Theorem:**

If a variable X has a distribution with a mean (μ) and a standard deviation sigma (σ), then the sampling distribution of the mean, based on random samples of size N, will have a mean equal to μ and a standard deviation

$$\sigma_{\bar{x}} = \frac{\sigma}{\sqrt{N}}$$

and will tend to be normal in form as the sample size increases.

A number of points need to be made about the Central Limit Theorem.

- First, the theorem describes the relationship between the sampling distribution mean and the population mean. It specifies that the mean of the sampling distribution will be equal to the mean of the population. This makes intuitive sense. If we selected all possible samples of a given size from a population, and if these samples were truly random, then the mean of all these samples should equal the population mean.

- Second, the Central Limit Theorem describes the relationship between the standard deviation of the population and the standard deviation of the sampling distribution. The formula shows that as the square root of the sample size increases, the standard deviation of the sampling distribution decreases. What this means is that as the sample gets larger, the sample does a better job of estimating the corresponding population parameter. The standard deviation of the sampling distribution is in fact a measure of random sampling error and is called the **standard error**.

- Third, the Central Limit Theorem states that the sampling distribution tends to be normally distributed if the sample is of sufficient size, regardless of the shape of the original population distribution. Of course, there are many possible shapes of population distributions, and these will be related to the size of the sample necessary to produce a normal or near normal sampling distribution. Under most circumstances, it appears that a sample size of 30 or more results in sampling distributions approximating normality.

Although the theorem presented above refers to the sampling distribution of the mean, similar principles apply to other sampling distributions, and *without some specification of an appropriate sampling distribution, an inferential statistical test is not possible.* The reason for this is that the characteristics of the sampling distribution are known and become the basis for making probability based statements about the population. In other words, they become the basis for statistical inference.

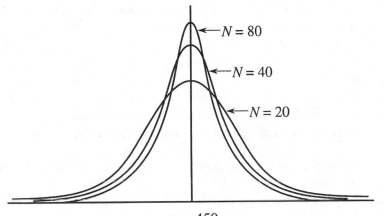

$$N = 80$$
$$N = 40$$
$$N = 20$$

$$\mu = 450$$

Figure 3.2. Sampling Distributions Based on Samples of 20, 40, and 80 Cases

? Some Examples: Sampling Distributions Based on Samples of Different Sizes

Imagine a normally distributed population distribution with a standard deviation of 50 and a mean of 450. Figure 3.2 shows three sampling distributions, based on samples of size 20, 40, and 80, drawn from this population. Note that, as specified by the Central Limit Theorem, all have a mean equal to the population mean of 450, and all are normally distributed around this mean. Applying the formula for the standard error to each sample size (i.e., 20, 40, and 80) produces three normal distributions, but as the sample size increases, the standard errors decrease and the distributions become clustered more closely around their mean of 450. Figure 3.2 demonstrates, without the use of more complex mathematical proofs, the critical relationship between the size of the sample and the precision with which the sample statistic accurately represents the population parameter. As sample size increases, sample statistics become tightly clustered around the parameters they estimate, with a range of error represented by the standard error of the sampling distribution in question.

The sampling distribution, though illustrated through the use of empirical examples such as those above, is a **theoretical probability distribution** that provides the information necessary for making inferences. In actual practice, only a single sample is selected, and this sample becomes the basis for making inferences concerning the population. These inferences may involve the construction of an interval around a sample estimate (interval estimation) or may involve the testing of a hypothesis regarding a specific value of a population parameter (hypothesis testing, or point estimation). Whether we are testing statistical hypotheses or constructing confidence intervals, it is our knowledge of the appropriate sampling distribution that provides the underlying theoretical support for our inferences from the sample to the population.

To summarize this discussion, let's review the main points.

1. Statistical inference involves generalization from sample statistics to population parameters.
2. To conduct inferential analysis, we must have a theory that underlies the process. The theory is based on probability.
3. There are two kinds of error in samples, random sampling error and bias. Through the selection of a random sample, bias can be eliminated and random sampling error can be measured.
4. Sampling distributions are theoretical probability distributions that describe the relationship between populations and samples.
5. The standard deviation of the sampling distribution is called the "standard error" and, when based on sample statistics, estimates random sampling error.
6. As the size of the sample increases, sampling error and the standard error will decrease.
7. The standard error is used both to develop interval estimates of population parameters and to conduct hypothesis testing.

In the two sections that follow, we briefly discuss confidence intervals and hypothesis testing.

WHAT IS A CONFIDENCE INTERVAL?

Consider the problem of attempting to estimate the population mean based on the sample mean (or a population proportion based on a sample proportion). In both situations, we recognize that the sample statistic we calculate from the sample data estimates the population parameter, but that there is likely to be some error in these estimates. In other words, the sample statistic is not likely to be *exactly* equal to the population parameter. On the basis of known characteristics of the distributions in question, and on our calculation of the standard error of the statistic we are using as an estimate, we can place an interval around a sample statistic that specifies the likely range within which the population parameter is likely to fall. This interval is called a **confidence interval**. The term *confidence interval* refers to the degree of confidence, expressed as a percentage, that the interval contains the population mean (or proportion), and for which we have an estimate calculated from our sample data.

Because the basis of statistical inference is probability theory, the estimates include a given probability of being incorrect. A 99% confidence interval will contain the population parameter 99% of the time, and a 95% confidence interval will contain the population parameter 95% of the time. More precisely, we construct an interval around a sample statistic. The population parameter will be contained within 99% of the intervals that are so constructed (or 95%). Thus, we know the likelihood that the interval will contain the population parameter of interest, and we also know the probability that the specified interval is incorrect.

If we want to satisfy the two criteria of both accuracy and precision, then a small standard error is required. Examination of formulas for standard error indicates that one obtains a small standard error by (a) reducing the variability in the sample, usually expressed as a standard deviation, or (b) increasing the size of the sample (making N larger). Because the values of the statistics contained in (a) are not directly under the control of the researcher, the most efficient way to achieve both accuracy and precision is to select a large sample. Sampling strategies used by professional researchers use knowledge of these relationships to determine in advance how large a sample must be selected to achieve the desired levels of accuracy and precision required by those who utilize the statistical information.

Of course, we don't know exactly how far a sample mean or proportion (the point estimate) is from the population value, any more than we can know how a horse's performance in a single race represents the horse's performance over a lifetime of races. A small confidence interval reflects greater precision, but a larger interval will generate greater confidence. Thus, other things being equal, a 99% confidence interval will be larger than a 95% confidence interval. Other things being equal, a large sample size generally produces a smaller interval around the sample value for a given level of confidence. This is the same thing as saying that as the sample size increases, the sample value becomes a better estimate of the population value. Figure 3.3 depicts 10 confidence intervals, each with a sample mean, all arranged around a fixed population mean. Two of these confidence intervals (contained within ovals) do not contain the population mean, whereas the other eight do. The point is that confidence intervals are estimates of the range within which the population value is likely to fall. Some

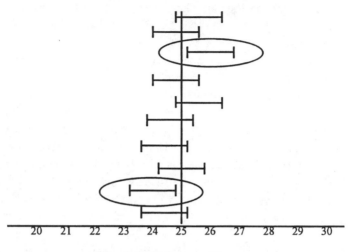

Population Value: $\mu_{\bar{x}} = 25$

Figure 3.3. Confidence Intervals Around a Population Mean (intervals within ovals exclude the population mean)

of these estimates will be incorrect, meaning that they do not contain the population value. With a 95% confidence interval, about 5% will erroneously exclude the population parameter; with a 99% interval, about 1% will do so.

To summarize, in order to produce a confidence interval you need to be able to calculate the standard error of the sampling distribution of the statistic of interest, typically means or proportions. The standard error of the sampling distribution can be reasonably approximated from the sample, if N is large enough. Practically speaking, an N of 50 generally is regarded as "large enough" (Bohrnstedt & Knoke, 1988). The formula for generating the upper and lower bounds of the confidence interval for a mean, using the t distribution, is

$$\bar{X} \pm tS_{\bar{x}}$$

where:

$\bar{X}$ = sample mean

t = the t value corresponding to the level of confidence desired, usually 95% or 99%

$S_{\bar{x}}$ = standard error of the mean, estimated from the sample standard deviation

WHAT IS A HYPOTHESIS TEST?

In addition to estimating the interval within which a population parameter is likely to fall, we may also test a hypothesis about a population parameter. A **hypothesis test** usually derives from a prior **research hypothesis** stating that a relationship exists between two or more variables, or that two or more groups will be different on some characteristic. For example, a research hypothesis suggesting that there will be gender differences on a measure of intimacy might be worded as follows: *Among couples married 5 years or more, males report higher levels of intimacy than females.* (In Chapter 6, we will discuss how one would determine the appropriate statistical test to use in this situation.) For now, let us assume that we will be testing the difference between the mean intimacy levels of males and females, and that all couples in our sample have been married 5 years or more. To test this research hypothesis statistically, we would formulate two **statistical hypotheses**. The first of these two statistical hypotheses, the **null hypothesis**, states that the mean for males on the intimacy measure *in the population from which the sample came* is equal to or less than the mean for females. Expressed in symbols:

$$H_0: \mu_m \leq \mu_f.$$

The second hypothesis, called the **alternative hypothesis**, states that the mean for males will be greater than the mean for females. Expressed symbolically:

$$H_1: \mu_m > \mu_f.$$

A number of points about the above two hypotheses are important. First, the two hypotheses represent what is called a **one-tailed test**. A one-tailed test is used because the research hypothesis specifies that males express more intimacy than females. Thus, we know the **direction** of the hypothesized difference (i.e., males greater than females). A **two-tailed test** would hypothesize that the means were different, but would not specify the direction of that difference. Second, the two hypotheses are mutually exclusive: They cannot both be true. Third, by *rejecting the null hypothesis* we support the alternative hypothesis and we support the research hypothesis.

As a researcher performing a statistical test, you must ultimately make one of two decisions relative to the null hypothesis.

1. **Reject the null hypothesis:** You conclude that the null hypothesis is false. This means that the alternative hypothesis represents the correct state of affairs in the population (i.e., in the population, men on average report higher levels of intimacy).

2. **Fail to reject the null hypothesis:** You conclude that the null hypothesis cannot be rejected. There is insufficient evidence to support the argument that males report more intimacy than females.

The logic of the above approach is often difficult to grasp. It is based on the fact that our goal is to disprove (i.e., reject) an idea because the probability of its being true is unacceptably small. This decision is made relative to the null hypothesis. If the decision is to reject this hypothesis, then the researcher concludes that the alternative hypothesis is supported, and in conjunction with this decision, the original research hypothesis receives support. This decision, however, always carries with it a level of risk, a certain probability that in fact we have made a mistake. This level of risk is expressed as the **level of significance** *of the test. As a researcher, you are able to specify the level of risk you are willing to accept. You do this by stating the level of significance at which you will perform a given test of statistical inference.*

Why don't we *accept* the null hypothesis? Accepting the null hypothesis would mean that we had *proven* that the population means were equal or that the mean for males was less than the mean for females. This might be the case, but unless we had sampled all cases from a known population (in which case statistical inference would no longer be necessary), we would not be certain of this conclusion. Even if the two sample means were identical, we could not "accept" the null hypothesis because we cannot rule out the fact that we could have obtained identical sample means from populations with different means.

This reasoning has an interesting analogy in the legal system. In the American legal system, the burden of proof lies with the prosecution. The presumption is that the defendant is innocent. In statistical language, this represents the assumption that the null hypothesis is true. Although a jury must be convinced "beyond a reasonable doubt" that the defendant is guilty before it rejects the presumption of innocence (i.e., the decision to convict is made), the scientific community must be convinced, at a given level of risk, that variables are related (or the groups are different in the population) before the decision to reject the null hypothesis is made. The level of risk or **level of significance** is usually predetermined to be .05 or .01, meaning that we will reject the null hypothesis only when the probability of falsely rejecting the null hypothesis is less than 5 in 100 (.05) or less than 1 in 100 (.01). When we reject the null hypothesis, the results of the study are said to be "statistically significant" at a predetermined level of .05 or .01.

Historically, researchers have been preoccupied with the concept of statistical significance and usually have based the success or failure, as well as the publishability, of their studies on their ability to achieve it. Convention suggests that the level of significance should be set at a value of .05, .01, or perhaps .001. (The level of significance often is called a "p" value because it represents a probability, and typically it is italicized or underlined in published work.) Note that there is nothing particularly sacred about these values, and our reliance on them can be attributed to historical coincidence. The .05 p value has been selected as a good decision-making standard because it is sufficiently stringent to safeguard against accepting too many insignificant results as significant, while also not being too difficult to achieve. Those of us who grew up in a pre-computer age were instructed to use tables in the appendix of a statistics book to determine if our test statistics (e.g., χ^2, t, F) were "significant" or not. The tables generally were reserved for p values of .05, .01, or .001. The enterprising researcher would scan the tables to determine what test value would be necessary to achieve, for example, .05 significance with a certain number of degrees of freedom (df). The use of alternative values would have necessitated cumbersome statistical computations. Today, some statistics texts dispense with these tables, simply because available computer programs can easily generate the precise significance level (i.e., p value) associated with a given test. Many journals now require authors to specify the precise significance levels associated with their findings rather than merely indicate whether or not the results are "significant." (We support this practice because it provides more information to the reader, helps diffuse the mythical sanctity of the fixed alpha approach, and facilitates meta-analyses [see Chapter 11].)

In sum, the above reasoning forms the logical basis for what has become the foundation of statistical analyses for the past 50 years. It has served to define research as valid or invalid, meaningful or inconsequential, and worth publishing or unpublishable. In and of itself, there is little wrong with the logic of the system described above; however, as will be described in the following chapter, these principles have been misunderstood and misapplied. Statistical inference, which has served as the conventional standard judging the validity of social science research, has lost much of its credibility among contemporary researchers.

CHAPTER four

Rethinking Traditional Paradigms: Power, Effect Size, and Hypothesis Testing Alternatives

What is . . . ? What are . . . ?

How do I . . . ? When should or shouldn't I . . . ?

Be more concerned with effect size than significance testing?

Calculate an effect size?

Calculate power?

Calculate the confidence interval?

Determine an appropriate sample size?

Increase design sensitivity?

Increase the power of a study?

Interpret an effect size?

Report *p* values?

Use confidence intervals instead of *p* values?

Use confidence intervals in reporting results of statistical analyses?

A counternull value?

A reasonable sample size?

Alpha error?

Alpha level?

An odds ratio?

Bayesian statistics?

Beta error?

Design sensitivity?

Effect size?

Errors in hypothesis testing?

Inverse probability error?

Statistical power?

The binomial effect size display?

The coefficient of determination?

The critiques of hypothesis testing?

The critiques of statistical inference?

The difference between statistical and substantive significance?

The meaning of "alternative hypothesis"?

The meaning of "null hypothesis"?

The meaning of "statistically significant"?

The problems with traditional statistical testing?

The relationship between sample size and power?

The relationship between standard error, power, effect size, and sample size?

The relative importance of effect size and statistical significance?

Type I and Type II error?

The different effect size indicators?

Effect sizes associated with different statistical tests?

Level: Intermediate/Advanced

Focus: Conceptual/Instructional

four

Rethinking Traditional Paradigms

Power, Effect Size, and Hypothesis Testing Alternatives

● ◠ Chapter 3 introduced the nature of statistical significance testing in the social sciences. This chapter serves as a counterpoint to that discussion and introduces contemporary developments in statistical thinking. Whereas many readers no doubt experienced the previous chapter as a redux of material that has been presented in some form in almost all elementary statistics courses, this chapter contains both extensions and challenges to traditional thinking about statistics. Most of these ideas are not new, but traditional beliefs about hypothesis testing are so well ingrained in the social science community that divergent perspectives have not yet made a substantial impact on research practice.

WHAT IS THE DIFFERENCE BETWEEN STATISTICAL AND SUBSTANTIVE SIGNIFICANCE?

One of the most unfortunate sources of confusion in the use of statistical tests concerns use of the word "significant." Statistical significance and **substantive**

significance are not the same thing; in other words, a finding of statistical significance need not be a substantively important one. For example, consistent changes of a few points on a depression scale may or may not be indicative of significant affective and behavioral changes in the real world. The relative meaning of the term "significant" is exemplified by the difference in speed between 65 miles per hour and 75 miles per hour compared to the difference between 45 miles per hour and 55 miles per hour. The differences are statistically equivalent, but the meaning of the differences is substantially different for a driver when the speed limit is 65 miles per hour. The meanings we assign to scores cannot be dictated by statistical rules.

One can argue that statistical significance is a necessary but not a sufficient condition for substantive significance (Gold, 1969), but even that argument may be suspect when one considers the likelihood of making errors of inference in studies with small sample sizes. At best, significance tests offer a formal decision-making technique for identifying systematic covariation in a set of data. Here are the words of William Hays (1993), author of a perennially best-selling statistics text:

> All that a significant result implies is that one has observed something relatively unlikely given the hypothetical situation. Everything else is a matter of what one does with this information. Statistical significance is a statement about the likelihood of the observed result, nothing else. It does not guarantee that something important, or even meaningful, has been found. (p. 384)

In sum, statistical significance can be thought of in two related ways. First, it may be thought of as the ability to *generalize.* When the statistical test is "statistically significant," we place confidence in the decision to generalize the findings from the sample to the population. Second, significance may be thought of as the *probability of making an error.* When one rejects the null hypothesis, the conclusion is that the null hypothesis is false; however, because this decision is based on a probability, there is also a probability that the null hypothesis might be true, or at least that the differences or relationships we are examining are so small as to be functionally equivalent to zero. Thus, if we reject the null hypothesis under these conditions, an error has been made. The probability of making this type of error, called a **Type I error**, is given by the level of significance (i.e., the p value) of the test. It is also possible to make a second type of error, called a **Type II error**. A Type II error occurs when the researcher fails to reject the null hypothesis when it is false. There is considerable debate

TABLE 4.1 Errors in Hypothesis Testing

Decision	Null Hypothesis Is	
	True	*False*
Reject null	Type I Error (Alpha [α] error)	Correct decision ($p = 1 - \beta$)
Fail to reject null	Correct decision ($p = 1 - \alpha$)	Type II Error (Beta [β] error)

about the merits of a strong emphasis on Type I errors in scientific research, whereas Type II errors are often ignored. One result of this has been that Type II errors probably are more common than Type I errors, simply because their probability is not acknowledged. We have quite a bit to say about this in the following sections.

WHAT IS STATISTICAL POWER?

Nearly every statistics text we have encountered contains some version of Table 4.1. This table shows the relationship between the researcher's null and alternative hypotheses, and the "true" state of affairs of the population. Because the null hypothesis can be either true or false, and the researcher can either reject or fail to reject this hypothesis, there are two ways to make correct decisions and two ways to make errors.

We have noted that **Type I error** consists of erroneously rejecting a "true" null hypothesis, that is, concluding that there is a relationship between variables, or a difference between groups, when in fact there is not (that is, when the sample results reflect a difference that is not representative of the population as specified by the null hypothesis). Demonstrations of probability theory that underlie these considerations often draw on the example of flipping coins: How many consecutive heads or tails would you insist on before concluding that you have a coin with either heads or tails on both sides? Researchers in the social and behavioral sciences usually have opted for conservative decision making, preferring to ignore differences or relationships that may exist rather than erroneously assert differences or relationships that do not exist. Another way of saying this is that most researchers would rather be blind than gullible (Axinn, 1966). They set the chances of making a Type I error to no more than .05 (also called **alpha**), meaning that they would be willing to be wrong no more than 5% of the time in claiming a relationship between variables that in reality does not exist in the population. Of course, investigators would also like to limit Type II errors (called **beta** errors) and correctly discern relationships between variables that do exist by rejecting a null hypothesis. Note that preferences with regard to making Type I errors as opposed to Type II errors are value judgments and depend on the implications of the conclusions. American society encourages juries to find defendants not guilty (i.e., fail to reject the null hypothesis of innocence) if there is reasonable doubt, on the assumption that it is preferable to allow a criminal to evade justice than to convict an innocent party.

Statistical power refers to the ability of a statistical test to detect relationships between variables. Another way of saying this is that power refers to the probability of rejecting the null hypothesis when it is false, and, therefore, should be rejected. Because the probability of failing to reject the null hypothesis when it is false is referred to as beta error (see Table 4.1), power can be described as $1 - \beta$. Thus, if for any given test the probability of a Type II error [β] was .2, the power of the test would be .8, which means that 80% of the time the investigator will find an effect that exists in the population. In practice, .8 seems to be the minimum level of power that researchers deem acceptable to conduct meaningful statistical analyses.

The concept of power comes from a 1933 paper by Jerzy Neyman and Egon Pearson. Jacob Cohen (1962, 1977, 1988) is generally considered the father of modern power analysis. Despite Cohen's cogent arguments and the acknowledged need to increase the power of empirical studies, there has been no discernible increase in the power of published studies since Cohen first published a power analysis of studies in 1962 (Sedlmeier & Gigerenzer, 1989). This state of affairs probably results from some commonly held misconceptions about the attributes of significance testing as well as the glacial speed at which innovation in data analysis usually takes a foothold in the social sciences. Whereas R. A. Fisher, the early voice of null hypothesis testing, once compared statisticians concerned with Type II error to "Russians" trained in technological efficiency instead of scientific inference, almost everyone agrees today that consideration of power is indispensable to decision making based on statistical inference (Sedlmeier & Gigerenzer, 1989).

Comprehending the concept of power is not easy, and an example might help. Let's imagine that your research hypothesis suggests that a relationship between two variables exists. We will call these two variables X and Y. Let's also assume that your hypothesis is correct, and that *in the population from which you will sample*, the correlation between these two variables is .25 (ρ = .25). You set the alpha level (i.e., level of significance) at .05 and draw a random sample of 15 cases from this population. You obtain an observed correlation of .29 (r = .29). The difference between the .25 in the population and the .29 in the sample represents random sampling error. In other words, in this example we are assuming that random sampling error produced a sample in which the correlation is slightly larger than the correlation in the population as a whole. We have used a Greek letter to represent the population correlation (ρ) and an Arabic letter to represent the sample correlation (r). Using standard statistical practice, you would test the null hypothesis that ρ = 0. Despite the fact that your research hypothesis is correct, under these conditions you would fail to reject the null hypothesis and would make a Type II error (i.e., fail to reject a false null hypothesis). What exactly is the probability of a Type II error (β) given this scenario? The answer is .85, meaning that the power level is .15, woefully below the minimum level of .80 we recommended earlier. In fact, to have made the correct decision in this situation (i.e., to reject the null hypothesis at an alpha level of .05), you would have had to obtain an observed correlation of .512 or larger. The critical point here is that *small sample sizes effectively reduce power and may make it very difficult to achieve statistical significance at even the .05 level*. Sadly, the literature offers copious evidence of how research studies in the social sciences typically are underpowered for detecting all but very large effects (Lipsey, 1990; Reed & Slachert, 1981; Sedlmeier & Gigerenzer, 1989).

So how could we have increased the power of the above study? Power is a direct function of four variables: (a) alpha level, (b) sample size, (c) effect size, and (d) the type of statistical test being conducted. Because we typically consider power within the context of a given statistical test, the type of test being conducted becomes an important consideration. Different tests may be more likely to detect an effect given the characteristics of the data in question and the type of effect, and, therefore, be more "powerful." (The reader is referred to power tables for common parametric tests provided by Cohen [1977, 1988].)

As previously suggested, an alpha level of .05 is fairly standard in the social sciences. Even though the beta level decreases as the alpha level increases (e.g., from .05 to .10), thus increasing power, most social scientists would not be willing to increase the probability of a Type I error from .05 to .1 even if this represented a substantial increase in power.

Effect size, the next variable influencing power, is a function of the actual size of the relationship in the population under investigation. Because, for the purposes of this example, we have set this value at .25, and this value generally is not under the control of the researcher, only sample size remains as a

TABLE 4.2 Relationship Between Sample Size and Power

Population Correlation (ρ)	Alpha Level (α)	Sample Size (N)	Type II Error (β)	Power (1 – β)	Minimum Significant Sample Correlation (r)
.25	.05	15	.85	.15	.512
.25	.05	30	.73	.27	.360
.25	.05	60	.51	.49	.254
.25	.05	120	.21	.79	.179

reasonable candidate for modification. Table 4.2 shows the relationship between sample size, beta error, and power for the above example.

As shown in Table 4.2, one would need to double the sample size three times, to 120, to achieve a power level of approximately .79. With this many subjects, an observed correlation of .179 would be statistically significant at the .05 level of significance. This value (.179) is obtained from the "Minimum Significant Sample Correlation" column of Table 4.2 and shows the smallest observed value that would prevent a Type II error. Another way to say this is that if the population correlation is .25, there is a 79% chance that the sample r value will be .179 or larger, and a 21% chance that the r value will be less than .179 and not statistically significant, resulting in a Type II error. Clearly, with only 15 or 30 subjects, the observed correlation in the sample must be much larger than the true value of the population correlation hypothesized here to avoid a Type II error. It is thus clear that power depends heavily on selecting a sample size sufficient to detect the relationship in question.

Examination of Table 4.2 leads to the question, "What would happen to the other values in the table if the value of the population correlation were larger (or smaller), thus increasing (or decreasing) the size of the relationship in the population (i.e., the effect size)?" It makes intuitive sense that if the population correlation is large (a strong relationship), that value should be easier to detect in a sample than if the value is small (a weak relationship). The strength of a relationship between two or more variables in the population is referred to as an **effect size**, because the effect of one variable on another may represent a relationship that is strong, weak, or somewhere in between. For example, given our hypothetical model in which $N = 15$ and alpha = .05, the required population correlation (ρ) to achieve a power level of .80 is .66. Said another way, with a population effect size of .25, the power level is only .15, but with the relatively much stronger effect size of .66, power increases to .80. Most contemporary journal articles will highlight the relative importance of statistical significance over and above a consideration of the size of the effect. This raises questions of whether an effect size is as important as a significance level and how effect size should be reported and used. The answers to these questions invite controversy and deserve consideration.

Cohen has suggested that an effect size index (referred to as Cohen's d) be constructed based on the difference between the experimental and control group means, expressed in standard deviation units. Such a measure serves as an adjustment for differences in scales and permits comparisons across many studies, such as those used in meta-analysis (see Chapter 11). A simple formula for the population effect size in a two-group experimental design (containing an experimental and a control group) would be

$$\frac{\mu_E - \mu_C}{\sigma}.$$

When estimating the population effect size using sample statistics, the formula would be

$$\text{Sample Effect Size (ES')} = \frac{\overline{X}_E - \overline{X}_C}{S}.$$

For example, a two-group experimental design with means of 50 in the experimental group and 44 in the control group, and a common standard deviation of 8, would have an effect size of $(50 - 44)/8 = .75$. An effect size of .75 is considered large, using guidelines suggested by Cohen. As the within-group variability increases, represented by an increase in the common standard deviation, the estimate of effect size decreases. For example, if the common standard deviation in the above example were doubled, to 16, the effect size would be cut in half, to .375. This value would be considered in the small to medium range by Cohen. In this context, Cohen suggests that an effect size of .2 is "small," .5 is "medium," and .8 is "large." The relationship between the effect size and the t ratio is given by the formula

$$t = \frac{ES}{\sqrt{\dfrac{1}{n_1} + \dfrac{1}{n_2}}} \quad .$$

Thus, in a two-group experiment with 20 cases per group and an effect size of .75, the t value would be

$$t = \frac{.75}{\sqrt{\dfrac{1}{20} + \dfrac{1}{20}}} = 2.37.$$

With 38 degrees of freedom $(20 + 20 - 2 = 38)$, this value is statistically significant at alpha $< .05$, but not alpha $< .01$. With a larger standard deviation and smaller effect size, as suggested above, the t value would be

$$t = \frac{.375}{\sqrt{\dfrac{1}{20} + \dfrac{1}{20}}} = 1.19.$$

This value is not statistically significant at alpha $= .05$.

What about studies that do not have mean differences to compare, such as studies utilizing correlation and regression techniques? Cohen (1988) suggests that correlations of .50 represent large effects, .10 represent small effects, and .30 represent moderate effects. Cohen (1988) points out that tables describing the relationship between sample size, effect size, and power (i.e., power tables) typically assess t tests by referring to the sample size in each group, whereas those assessing power for correlations refer to total sample size. Thus, the *total* sample size required for r usually is smaller than that required for a t test.

Numerous measures of effect size may prove useful in statistical studies, and the measure utilized often depends on the type of statistical analysis being conducted. Excellent summaries of categories of effect size are found in recent articles by Kirk (1996) and Snyder and Lawson (1993). There are two basic classes of effect sizes: variance accounted for measures and standardized difference measures. Standardized difference measures *directly* examine differences between means and include Cohen's d and Hedges's g. Variance accounted for measures describe how much of the variability in the dependent variable is associated with variability in the independent variable(s) and range in value from 0 to 1. The bivariate correlation coefficient is the most popular variance accounted for measure. In addition to the correlation coefficient, there is also the "percent variability explained" by a statistical relationship, represented as the square of the correlation coefficient (r^2) and the square of the multiple correlation (R^2) typically encountered in multiple regression analysis. When conducting analysis of variance, **eta** may be used as a measure of effect size, and the percent variability explained may be represented by **eta**^{2} in a manner analogous to the use of r and r^2. Eta can be represented as

$$\sqrt{\frac{SS_{between}}{SS_{between} + SS_{within}}}.$$

In other words, eta is the square root of the proportion of variance in the dependent variable accounted for by the factors in a factorial design. Eta can be interpreted as a general case of r, the Pearson product-moment correlation coefficient, because it is nonspecific and may refer to non-linear as well as linear relationships.

Table 4.3 summarizes small, medium, and large effect sizes for commonly employed statistical techniques. The table can serve as a shortcut to answering the statistical consultant's most frequently asked question: "How many subjects do I need?" Realize, however, that all effect size estimates have certain constraints, and they must be interpreted within the context of an area of inquiry. Uncorrected effect sizes such as R^2 and eta^2, for example, overestimate population effects because they capitalize on the variance of the study sample. (Chapter 10 contains a discussion of this point in the context of regression analysis.) This positive bias, and the recommended statistical corrections for bias, are exacerbated with small sample sizes (Thompson, 1993). Zwick (1997) cautions, as well, that a large effect size does not guarantee that the results are interesting or practically significant any more than a low p value offers that assurance. Zwick also warns us not to rigidly adopt Cohen's definitions of small, medium, and large effect sizes, but to think for ourselves within the context of an area of study.

To reiterate, the **effect size** refers to any of several measures of the strength of a relationship. The simplest measure of association that conveys information about the degree of relationship between variables is the correlation coefficient, and we have used the correlation in our initial examples above. Thus, the most straightforward method of computing effect size is simply to compute the product-moment correlation and use this measure as is, or convert it into an effect size index using the formula provided earlier.

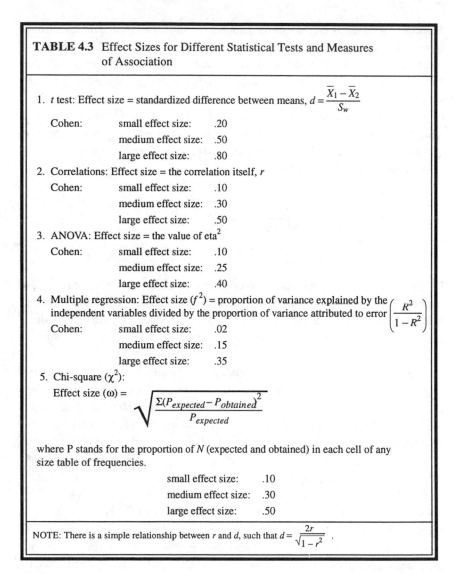

TABLE 4.3 Effect Sizes for Different Statistical Tests and Measures of Association

1. t test: Effect size = standardized difference between means, $d = \dfrac{\overline{X}_1 - \overline{X}_2}{S_w}$

   Cohen: small effect size: .20
             medium effect size: .50
             large effect size: .80

2. Correlations: Effect size = the correlation itself, r

   Cohen: small effect size: .10
             medium effect size: .30
             large effect size: .50

3. ANOVA: Effect size = the value of eta^2

   Cohen: small effect size: .10
             medium effect size: .25
             large effect size: .40

4. Multiple regression: Effect size (f^2) = proportion of variance explained by the independent variables divided by the proportion of variance attributed to error $\left(\dfrac{R^2}{1 - R^2} \right)$

   Cohen: small effect size: .02
             medium effect size: .15
             large effect size: .35

5. Chi-square (χ^2):

   Effect size (ω) = $\sqrt{\dfrac{\Sigma(P_{expected} - P_{obtained})^2}{P_{expected}}}$

   where P stands for the proportion of N (expected and obtained) in each cell of any size table of frequencies.

             small effect size: .10
             medium effect size: .30
             large effect size: .50

NOTE: There is a simple relationship between r and d, such that $d = \dfrac{2r}{\sqrt{1 - r^2}}$.

How large an effect should a researcher seek to detect? Many researchers seem to opt for being able to identify medium effects. Given that a researcher knows the effect size that he or she considers important to detect, the type of statistical test and alpha level that will be used, and the desired power, it then becomes possible to calculate the appropriate sample size, using hand calculations, tables, or computer programs that provide the researcher with recommended sample sizes. For example, conducting research that requires a bivariate correlation analysis, with an alpha of .05, a power level of .80, and an assumed medium effect size of $\rho = .30$, yields a required sample size of 82. The point of this example is that once the effect size is estimated, an alpha level is set, and the statistical test is chosen, the size of the sample necessary to detect this effect is almost a given. A researcher can easily calculate how much smaller an effect could be detected by increasing the sample size in increments or, conversely, decreasing the effect size in increments and determining how much larger the sample must be to uncover this effect. Hair, Anderson, Tatham, and Black (1995) state that power can be problematic with $N < 50$ and goes up slowly with more subjects when N becomes greater than 150. As sample sizes increase, any small difference becomes statistically significant, and it becomes critical to consider the practical significance of the findings by examining the size of the effect as well as its statistical significance.

HOW DO I INTERPRET AN EFFECT SIZE?

Some statistical analyses lend themselves to correlation-based indices of effect size such as r or r^2. Interpreting these indices can be a little tricky. In most contexts, it is the square of the correlation coefficient that describes the explained variance two variables have in common. Thus, a correlation of .12 might be statistically significant with a fairly large sample size, but it implies shared variance between the variables of only 1.44% ($.12 \times .12 \times 100$), a relatively weak association. Whether or not a researcher assigns importance to this value will depend on the nature of the problem. The r^2 measure, which is sometimes known as the **coefficient of determination**, is not an easy concept to grasp intuitively.

Some authors have noted that r^2 may underestimate the practical effects of small influences that sum to produce meaningful results (Abelson, 1985). Following is an example where the proportion of variance explained (r^2) is an underestimate of predicting an outcome variable from a predictor variable (Abelson, 1985). Imagine two groups of baseball players, one batting .222, one .333. Anyone who follows baseball will view this performance disparity as substantively significant. But wait a moment. The correlation between the manager's choice of batter and getting a hit is only .111. Thus, the manager controls only .0123 or 1.23% of the variance in the outcome ($.111^2 = .0123$). If the manager chooses the player with the higher batting average, the predicted outcome is improved by 11 percentage points ($.333 - .222 = .111$), an increase

TABLE 4.4 Crosstabulation and Binomial Effect Size Display for Effect of Weight Loss Program

Weight Loss	Crosstabulation			Binomial Effect Size Display		
	Control	Treatment	Total	Control	Treatment	Total
Lost weight	50	98	148	46.5	53.5	100
No change	69	101	170	53.5	46.5	100
Total	119	199	318	100.0	100.0	200
$r = .07$ $(r^2 = .0049)$						

of 50% in the probability of obtaining a hit! The implication is that the correlation, r, is often a better index of the effect size than r^2, the proportion of variance in Y explained by X. As in reliability estimates, in genetics research in intelligence, and in prediction studies, r can also be the best indicator of the proportion of variance in some variable Z, common to X and Y. The reader is referred to Ozer (1985) for additional arguments for the use of the correlation coefficient rather than the squared correlation as a meaningful estimate of practical effect size.

Rosenthal and his associates (Rosenthal & Rubin, 1979, 1982; Rosnow & Rosenthal, 1996) have suggested that a binomial effect size display (BESD) may make more intuitive sense in evaluating the practical utility of various correlation coefficients. This display helps us visualize the effect of a new treatment intervention, selection method, or predictor variable on some measure of success rate, such as improvement, cure, or survival. Outcome values are transformed into uniform percentages and presented as a 2×2 contingency table with marginal values set at 100%. The columns of the table represent the independent variable (treatment vs. control), and the rows of the table represent the dependent variable as a dichotomous outcome. For example, the outcome could represent distinctions such as sick vs. healthy, chosen vs. nonchosen, or success vs. failure. Table 4.4 illustrates this point for 318 persons attempting to lose weight, half of whom listened to motivational tape recordings (treatment) and half of whom did not (control).

Note that this example makes use of a dichotomous outcome variable. Studies in which the relevant correlation is based on a continuously distributed outcome variable can be similarly transformed to a binomial effect size display (Rosnow & Rosenthal, 1996). The change in improvement from the treatment group (calculated as $.50 + r/2 \times 100$) to the control group (calculated as $.50 -$

$r/2\times 100$) is equivalent to a percentage increase in improvement resulting from the treatment. This value is shown in the crosstabulation table that presents the raw data with column percentages. Practically speaking, this rate of improvement may be substantial and is certainly more impressive than the shared variance (percentage of variance explained) would imply, .0049 in this example.

The relationship between effect sizes and statistical significance is related to the relationship between power and statistical significance. One can readily obtain the same effect size, but nonsignificant results, with a smaller sample. Rosnow and Rosenthal (1989b) compare two researchers working on the same problem. Smith, using an $N = 80$, finds that leadership style A is better than leadership style B. He bases his conclusion on a t value of 2.21, $df = 78$, $p = .05$. Jones, the inventor of leadership style B, uses only 20 subjects and cannot replicate Smith's findings ($t = 1.06$, $df = 18$, $p = .30$), yet both studies have identical effect sizes ($r = .24$). The difference lies in the power of the studies: Smith, with 80 subjects, had the power (.60) to reject the null hypothesis (at alpha = .05), compared to Jones's power of .18 with 20 subjects.

The same authors (Rosnow & Rosenthal, 1989b) provide a good example of the limitations of significance testing without the presence of a measure of effect size. In 1988, the Steering Committee of the Physicians' Health Study Research Group reported that aspirin can effectively reduce the risk of heart attack. This conclusion was based on a 5-year study of a sample of 22,071 physicians. Approximately half of the sample was given 325 mg of aspirin every other day, while the other group was given a placebo. The argument is that aspirin promotes blood circulation and indirectly reduces mortality from myocardial infarction. Table 4.5 summarizes the data from the study.

Based on the outcome data, 1.3% of the physicians suffered a heart attack, broken down into 0.9% of the aspirin group and 1.7% of the placebo group. Even with this low number of heart attacks, the difference between groups is statistically significant well below customary alpha levels, $\chi^2 (1, N = 22,071) = 25.01$, $p < .00001$. This leads to a relatively confident conclusion that regular administration of low levels of aspirin can effectively reduce the rate of heart attack.

The lower part of the table describes the mortality of those subjects who did experience a myocardial infarction during the 5 years of the study. As predicted, 4.8% of those in the aspirin condition experienced a fatal heart attack, compared to 9.5% in the placebo condition. This difference, however, is not quite significant at the customary .05 level ($p = .08$). Certainly the difference in sample sizes between the two comparisons accounts for these different conclusions.

What happens if we take effect size into account? When the effect size is computed as a standard Pearson product-moment correlation in the first analysis,

TABLE 4.5 Aspirin's Effect on Heart Attack Prevalence

Condition	No Heart Attack	Heart Attack
Aspirin	10,933	104
Placebo	10,845	189

	Lived	Died
Aspirin	99	5
Placebo	171	18

SOURCE: Adapted from Rosnow and Rosenthal (1989b).

we get $r = .034$, which at first glance suggests a very unimpressive, minuscule relationship between aspirin and heart attack likelihood. As in Abelson's (1985) baseball example, however, small effect sizes can be very important when measurement error is small. We can take this correlation of .034 and transform it into binomial language. This transformation leads to the outcome that 3.4% fewer subjects would be likely to experience a heart attack if they took the aspirin regimen. This is not an inconsequential conclusion, given life and death outcomes. It also suggests that before concluding that aspirin does not reduce fatal heart attacks (remember the $p = .08$), a larger group of subjects probably should be sampled.

? An Example Using Odds Ratios

An alternative approach to examining the data from the aforementioned aspirin study makes use of the so-called odds ratio. The **odds ratio** is a measure of association likely to be encountered in research studies with dichotomous outcomes, such as contracting or not contracting a disease, and in which we can determine whether or not the case (usually a person) experienced something we believe to be associated with the outcome. For example, Table 4.6 reproduces the aspirin and heart attack data presented in Table 4.5. The odds of a heart attack and the odds of dying in each group, aspirin vs. placebo, are obtained by dividing the number of persons who had a heart attack or the number of people who died in each group, by the number of people who did not. Thus, as shown in the table,

TABLE 4.6   Aspirin's Effect on Heart Attack Prevalence: Odds Ratio Analysis

Condition	No Heart Attack	Heart Attack	Odds of Heart Attack	Odds Ratio
Aspirin	10,933	104	.0095	1.83
Placebo	10,845	189	.0174	
Condition	Lived	Died	Odds of Dying	Odds Ratio
Aspirin	99	5	.0505	2.08
Placebo	171	18	.1053	

SOURCE: Computed from data in Rosnow and Rosenthal (1989b).

the odds of a heart attack for those who took aspirin are .0095 and for those who did not take aspirin are .0174. The ratio of the odds in the non-aspirin or placebo (called the unexposed) group to the odds in the aspirin group is called the "odds ratio." The odds ratio in this example is .0174/.0095 = 1.83. In other words, the risk of a heart attack is approximately 1.8 times as great among the non-aspirin group. Similarly, the odds ratio for death following a heart attack in the non-aspirin group, compared to the aspirin group, is 2.08. This means that death is twice as likely without aspirin.

The above example shows that if the odds were the same in the two groups, the odds ratio would be 1, and that odds greater than one indicate a greater likelihood of having a heart attack or dying, among the group that did not take aspirin. Said another way, odds ratios of one or close to one generally indicate that there is no relationship between exposure or intervention and outcome.

This lengthy discussion has a number of implications. One is the advisability of designing studies with sufficient power to find differences that exist. This is commonly referred to as **design sensitivity**. A sensitive design is one that has the ability to detect an effect when it is actually there. When the researcher knows the size of the effect he or she would like to detect, then the tools are available to design a study that can reliably detect this effect, if in fact it exists. Such a design is considered sensitive. Based on a discussion by Titus (1994) and others, some suggestions for increasing the sensitivity and power of a design include the following:

1. Design the study to minimize the size of the standard error. This can be accomplished through the use of measures with small standard deviations and through the use of a homogeneous sample.
2. Maintain uniformity in conducting the study.
3. Select measures that are as sensitive to the desired effect as possible.
4. Make differences between your conditions as large as can be defended.
5. Use strong and powerful independent variables.
6. Make use of multiple measures.
7. Use a within-groups design, if appropriate, because this reduces the error term.
8. Use equal ns and focused tests or contrast tests and covariates if appropriate.

In summary, power is enhanced by adopting good, reliable measures, an appropriate design, and a treatment that produces large effects. Practically speaking, if your results are not significant, but are in the predicted direction, you can determine how many more subjects are needed to reach statistical significance. Power and effect size always need to be approached in the context of the particular phenomenon you are investigating. The following questions are especially important:

1. What does the research literature in the area say about reasonable effect sizes?
2. What are typical scores on my dependent variables, based on normative data and other studies or similar populations?
3. Based on theory, research, and common sense, what is the smallest detectable effect of practical significance in this study?

Finally, it is a good idea to use power calculations twice, at both the beginning and culmination of a study. At the beginning of a study, these calculations help determine the appropriate sample size and acquaint you with the findings of others. Effect size, however, is a characteristic of relationship in a population, not a general characteristic of the relationship itself; thus, reexamination of the power actually achieved is critical to assessing the meaning of the findings.

WHAT ARE THE RELATIONSHIPS AMONG SAMPLE SIZE, SAMPLING ERROR, EFFECT SIZE, AND POWER?

The relationships among sample size, sampling error, effect size, and power can be summarized as follows:

1. As population effect size increases
    a. Power will increase for any given sample size,
    b. The sample size necessary to achieve statistical significance will decrease, and
    c. The probability of a Type I error will decrease.
2. As the sample size increases
    a. Power will increase,
    b. Sampling error will decrease, and
    c. The probability of a Type II error will decrease.
3. As sampling error decreases
    a. Power will increase for any given effect size,
    b. The sample size necessary to achieve statistical significance will decrease, and
    c. The probability of a Type II error will decrease.

? Confidence Intervals Revisited

A finding of statistical significance means only that the true population effect is probably not zero, but typically it is (mis)understood as indicating that the study is of substantive significance. A finding that a study is not statistically significant means that we have insufficient evidence to conclude an effect is present, but often it is (mis)understood to mean that the absence of an effect has been demonstrated. Confidence intervals, by contrast, focus on the magnitude of the effect and on the precision with which the effect is estimated. It is reasonable to ask, therefore, whether or not a confidence interval might avoid some of the problems associated with traditional significance tests. Oakes (1986) argues that a confidence interval be utilized in lieu of significance testing because it focuses on what the effect is, rather than what it is not. In addition, in virtually any case in which a test of significance is appropriate, confidence intervals, interpreted like a significance test, would provide the same and additional information. For example, imagine a study of high school students that examines the relationship between self-esteem and grade point average (GPA). If, in a study of 50 students, this relationship was .22, the result would not be reported as statistically significant. In fact, this relationship might be totally ignored or summarized with a statement such as "The relationship between self-esteem and grade point average was not statistically significant." Clearly, this statement provides very little information. As an alternative, let's examine the 95% confidence interval for the correlation between self-esteem and GPA. We use procedures described by Blalock (1979) and Fisher's r to z transformation. Essentially, the steps to produce the 95% confidence interval are:

1. Using a table, convert the observed r to Fisher's z,
2. Compute the standard error of z from the formula $((1/\sqrt{N-3})$ (1.96)
   (where N is the sample size and 1.96 is the z score value for the 95% confidence interval),
3. Compute the limits of the confidence interval by subtracting and adding the above quantity from Fisher's z, and
4. Convert the zs for the lower and upper limits back to r from the same table.

The resulting confidence interval is

$$C(-.06 \leq \rho \leq .47)95\%.$$

This confidence interval tells us as much as, and considerably more than, a significance test.

1. First, the interval tells us that the effect size might be negative, or might be as strong as +.47.
2. Second, because the interval includes the value of ρ under the null hypothesis (i.e., H_0: $\rho = 0$), we know that the value is not statistically significant at $\alpha = 1 - .95 = .05$.
3. Third, we have an indication of precision of the estimate. Because the range of the interval includes both positive and negative values ranging from $-.06$ to $+.47$, the estimate does not appear particularly precise.
4. Finally, we can use the width of the confidence interval as the determining factor in setting the sample size for a study (Cobb, 1985; Gordon, 1987; Oakes, 1986).

Using the same procedures (Blalock, 1979), the 95% confidence interval, given the same observed correlation ($r = .22$) but with $N = 103$, would be

$$C(.03 \leq \rho \leq .40)95\%.$$

Note that this confidence interval *would* be statistically significant at alpha = .05 because the interval does not include the value of 0, and that the interval has become more precise (i.e., the range of values within which the population correlation is likely to fall is smaller).

A confidence interval can also be fit around an odds ratio. In such a case, an interval containing the value of 1.00 would not be statistically significant. Let's consider three reports of confidence intervals fit around odds ratios, all significant at the .05 level (because the lower confidence limit exceeds 1.0):

1. Odds ratio of 1.2 (95% CI: 1.1 to 1.4)
2. Odds ratio of 4.0 (95% CI: 1.1 to 16.0)
3. Odds ratio of 4.0 (95% CI: 3.9 to 4.3)

In the first example, the observed effect is small, showing only a 20% increase in risk, and the estimate is precise, because the range of values is small. In the second example, the observed effect is substantial but imprecise, because the range of values is large. In the last example, there is a substantial and precise estimate of the population value. In other words, knowing the confidence interval tells us considerably more than simply knowing that all three ratios are significant at .05.

As you know (see Chapter 3), you need only to have an approximately normal distribution and to know the standard error of the sample to be able to generate a confidence interval. Our original example was based on determining a confidence interval around a product-moment correlation. Methods for deriving confidence intervals have been developed for most complex statistics. The reader is referred to Cohen (1990), Loftus and Masson (1994), and Serlin (1993) for examples and computational details. Reichardt and Gollob (1987) offer a suggestion concerning how to report results based on confidence intervals in articles submitted for publication. An example cited by Becker (1991) reads:

The mean difference in reported liking for school between first-grade girls ($M = 4.26$) and first-grade boys ($M = 2.53$) is 1.73 (.95 CI = 0.03, 3.43). Because the mean difference of zero falls outside the lower limit of 0.03, the associated p value is below the .05 level and we can reject the null hypothesis at that level. (p. 654)

In sum, the confidence interval avoids many of the problems inherent in simply reporting the statistical significance of a test statistic. If the lower bound of the confidence interval exceeds the value predicted by the null hypothesis (e.g., an odds ratio exceeding 1.0 or a correlation exceeding 0.0), we can conclude with the specified level of certainty that the effect is real (i.e., exists in the population). The confidence interval thus can provide a traditional test of the null hypothesis plus the following information:

1. An estimate of effect size that is separate and distinct from the precision of the estimate.
2. The value for the lower bound of this value in the population and a value for the upper bound of this value in the population.

WHAT ARE THE PROBLEMS WITH
STATISTICAL SIGNIFICANCE TESTING?

Statistical significance testing is deeply entrenched within the social science research community. It holds an almost monopolistic sway over the way in which quantitative methods are taught in colleges and universities, and it is difficult to publish research studies in scholarly journals without following the procedures mandated by the logic of significance tests. There seems, however, to be a change brewing in the research winds. An increasingly vocal group of critics is leading the way toward reform in the area of data analysis and the use of inferential statistics. Rothman's (1986) statement exemplifies this critical perspective:

> Testing for statistical significance today continues not on its merits as a methodological tool but on the momentum of tradition. Rather than serving as a thinker's tool, it has become for some a clumsy substitute for thought, subverting what should be a contemplative exercise into an algorithm prone to error. (p. 445)

It is not as if such opinions are new to the scene. Indeed, Bakan (1966), Labovitz (1970), Meehl (1967), Morrison and Henkel (1970), and Rozeboom (1960), among others, cogently summarized the primary arguments against the use of significance tests three decades ago. The difference is that more and more statisticians are becoming increasingly vociferous and determined in lending their voices to the opposition (cf. Cohen, 1994; Hunter, 1997; Loftus, 1991, 1996; Schmidt, 1996), while others have met the challenge and responded (cf. Chow, 1988; Cortina & Dunlap, 1997; Frick, 1996). What are the concerns being expressed, and how seriously should they be taken? Below we provide four sets of responses to these questions.

First, the distinction and confusion between statistical significance and substantive significance already has been discussed. This argument led to a recommendation for including effect sizes in the reporting of statistical results. We suggest that you always include both a discussion of the statistical significance of your findings and the effect size. This usually can be accomplished through the use of confidence intervals. Confidence intervals and p values can be combined in a study to increase clarity and acceptability by those who are traditionally trained.

We refer you also to a recent article by Rosnow and Rosenthal (1996) that is intended to help research consumers correctly interpret the results of published studies and incorporate effect sizes into their own studies. In particular, the authors have proposed a new statistic, the **counternull value** of an obtained effect, which refers to the "non-null magnitude of effect size that is supported by the same amount of evidence as is the null value of the effect size" (p. 333). In other words, if one is concerned about reducing Type II error with the same level of commitment that traditionalists have invested in reducing Type I error, there are ways to construct intervals of p values ranging from the null value to the counternull value that indicate whether conclusions of no effect might be mistaken. Thus, the potential practical significance of a finding can be estimated even when the null hypothesis is not rejected.

The technology of counternull statistics makes use of **contrasts**, which are specific predictions of interest among research data from multiple groups, and of correlational measures of effect size. When a researcher uses Cohen's d as an effect size measure, the counternull statistic is twice the effect size measure d (i.e., counternull = $2d$). For example, with a $d = .31$, evaluated at a p level of .09, the counternull value is .62. This is because traditional significance testing is biased toward the identification of Type I errors at the expense of admitting Type II errors. When a researcher uses the Pearson r as a measure of effect size, the counternull statistic is derived from the following formula:

$$r_{counternull} = \sqrt{(4r^2)/(1 + 3r)} .$$

Exactly the same logic applies as above in interpreting these results.

Second, a major error in reasoning that is frequently conveyed is the inference that the failure to reject the null hypothesis implies the acceptance of the null hypothesis. In reality, the logic of hypothesis testing revolves around testing a null hypothesis that is seldom true, because "true" means that there is no difference between groups in the population, an inequality that is rarely the case. To demonstrate this point empirically, Lipsey and Wilson (1993) surveyed 302 different meta-analyses with an average of 60 studies each (see Chapter 11) and found only 3 cases out of 302 that yielded an effect size of 0. As Gregory Loftus (1996) has colorfully put it, rejecting a null hypothesis is akin to rejecting the proposition that the moon is made of green cheese! Hence, the failure to reject the null hypothesis does not mean that there is *no* difference between groups on a variable; it means that whatever differences exist are too negligible to be identified by a statistical test at a suitable level of significance. Almost all relationships between variables can be made statistically significant by making N sufficiently large. Significant results tell us only if we have enough power to detect differences.

One specific context where this fallacious logic is often employed is in concluding that groups are equivalent at baseline based on a demonstration that they are not statistically different from one another (Dar, Serlin, & Omer, 1994). We recommend against adopting this fallacious measure of equivalence and suggest that covariate analyses are the preferred procedure for testing non–randomly selected groups.

Third, researchers are seduced into believing that small p values (e.g., $p \leq$ .001) imply "highly" significant findings, whereas larger p values (e.g., .054) imply "almost" significant findings or "marginally" significant findings. These conclusions reflect an error in reasoning. Significance levels do not measure the strength of a statistical association; they measure the probability of the results given the validity of the null hypothesis. One cannot conclude that findings are large or important based on the significance level. Include the significance levels, but be cautious in interpreting them.

Fourth, similarly, the p value of a study does not indicate the probability of successfully replicating the results of the study (Oakes, 1986). Chances of replicating results are a function of the power of the study, which is very different from the significance level (p), which is calculated on the sample and not the population. We do, however, strongly support the practice of replicating individual studies and adopting multiple sources of corroboration in order to increase confidence in the results and validate a theory.

Hendrick (1990) notes that some new studies are "exact" replications that mimic the previous study as closely as possible; some are "partial" replications that attempt to duplicate the study as much as possible but change some procedural variable, such as a measure of one of the variables; and some are "conceptual" replications that seek to maintain the basic intent of the original study but sharply alter the procedure or intervention. External replicability of a study using a new sample is always highly desirable. When that is not practical, internal replication using the same sample of subjects may be considered. Thompson (1993, 1994) describes and evaluates three different approaches to internal replication: **cross-validation**, splitting the sample into two roughly equal groups and comparing results between them; **jackknife** methods (cf. Crask & Perrault, 1977), performing separate analyses with groups of subjects progressively eliminated from the analysis one at a time and comparing all possible analyses with each other; and **bootstrap** methods (cf. Diaconis & Efron, 1983; Lunneborg, 1987, 1990), copying the data set on top of itself over and over again to create an infinitely large data set and drawing samples from this "mega file" and comparing their results.

HAS SIGNIFICANCE TESTING
OUTLIVED ITS USEFULNESS?

One of the most vociferous critics of statistical significance testing is Frank Schmidt (1992, 1996), who has raised fundamental concerns about its utility. Schmidt begins his argument by referring to the standard decision rule in the field: If the test statistic is significant, there is an effect; if the test statistic is not significant, there is no demonstrable effect. Historically, this rule has been implemented by attending carefully to Type I error, that is, keeping the alpha level small, which means, of course, that Type II error (beta) can drift rather high. How high? Let us turn to an example offered by Schmidt (1996) that illustrates the implications of the relationship between Type I and Type II errors.

A researcher is evaluating the impact of a specific dosage of a particular drug on learning. The actual effect of the drug is to increase the amount learned by one half of a standard deviation (a medium effect size of .50), which corresponds to the difference between the 50th and 69th percentiles in a normal distribution. Think of this effect size as a known parameter of the population. It implies that 69% of the experimental group would have greater learning than the mean of the control group. Were we to repeat this study many times over by drawing new samples of subjects, with (let's say) 15 subjects per group, the effect size would vary because of sampling error. Under the null hypothesis the mean value of the effect size would be 0, and it would take an effect size of .62 or larger to generate a statistically significant effect at the .05 level (one-tailed). To most researchers, suggests Schmidt, this analysis is straightforward: The probability of a Type I error is limited to .05 with this significance test.

We have presupposed, however, that the real population effect size is always .50, which means that a Type I error is not possible. The Type I error is zero, not 5%! That is, the null hypothesis here is false. One can't make a Type I error, only a Type II error. If the effect is indeed present and the null hypothesis is false, the only type of error that can result is Type II. But this is precisely the kind of error that is not controlled. Figure 4.1 shows a comparison between two distributions: the distribution of the null hypothesis mistakenly used by most researchers, and the distribution of effect sizes with a mean of .50 (Schmidt, 1996, p. 117). To reach significance with this distribution, we still need an effect size of .62 or larger (.05 level), but this magnitude of effect will be found by only 37% of our studies, not 95% of them. Statistical power for this study is .37, a moderate value with a Type II error rate of .63. A two-tailed analysis would contain even less power (.26) and a higher Type II error rate (74%).

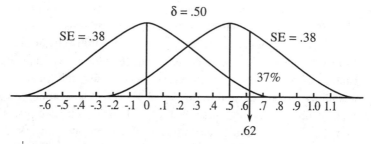

Figure 4.1. Statistical Power in a Series of Experiments
SOURCE: Adapted from Schmidt (1996, p. 117 [Figure 2]).

The point so far is that the potential error rate for a significance test is much higher than the .01 or .05 alpha level that is usually identified. If there is a treatment effect, the errors will be Type II errors, and these errors, maintain many critics, can be considerable, averaging perhaps 60% and potentially reaching into the .90s (Hunter, 1997). This is the ostensible reason for Cohen's repeated insistence over the past 30 years on conducting power analyses to reduce Type II error.

There are other peculiar and unfortunate ramifications of this study (Schmidt, 1996). The significance testing paradigm insists that an effect size must have a minimum value of .62 to be significant (most significant values will be higher), whereas the true effect size is always .50. A study that actually yields a correct effect size of .50 will be seen as nonsignificant! Given a large set of studies on this hypothesis, the majority (63%) will find no relationship between this drug and learning, and a minority (37%) will find a rather large effect (actually .89 in Schmidt's example, which is a much higher standard than the population effect of .50). It is difficulty in understanding the true meaning of these studies that has led Schmidt to promote meta-analysis as a superior approach to achieving cumulative knowledge in a field of study. More about meta-analysis can be found in Chapter 11.

Schmidt's (1996) arguments seem to be gaining increasing currency among statisticians in the social sciences. These issues, however, are complicated and controversial, and not every expert is ready to throw in the towel on significance testing. Cortina and Dunlap (1997), for instance, have marshaled a thoughtful analysis of the problems that have been raised and attempt to rebut many of them. They do not dispute the argument that, because the null hypothesis is always false, rejecting it cannot be an error, but they maintain that this fact has nothing to do with the Type I error rate, because Type I error has to do with a hypothetical distribution, not the actual sampling distribution of the test statistic. In other words, an alpha of .05 is the probability of making a Type I error *if* the null hypothesis were true, regardless of whether or not it actually *is* true.

Furthermore, Cortina and Dunlap (1997) point out that the term "null," in reference to the null hypothesis, is used to represent the hypothesis that is nullified (Cohen, 1994), which does not necessarily imply a hypothesis of no difference (that would be a "nil" hypothesis). One could just as easily select a null value of .10 and compare an experimental hypothesis to it. This is especially relevant to directional hypotheses because the rejection of a null hypothesis implies not only rejecting a specific difference (or lack of difference), but also rejection of all values at one end of a distribution. This is consistent with Frick's (1996) assertion that null hypothesis testing may not be ideal for making quantitative claims, but it is ideal for making so-called *ordinal claims*, claims that specify the order of conditions or effects or the direction of a correlation, rather than the size of the effect. In this regard, Tukey (1991) has contended that with *t* tests, the truth of the null hypothesis is not the central issue; the issue is the level of confidence we can have in stipulating the direction of the effect. Too often, perhaps, null hypothesis significance testing (NHST) is concerned with omnibus ANOVAs that deal only with the equality of means. Drawing conclusions about the direction of an effect is not a trivial matter, especially in fields such as psychology that do not easily generate *point* hypotheses, in contrast to studies in "hard" sciences such as physics, which rely on validating point hypotheses. Take, for instance, a physics demonstration that gravitational forces at sea level occur at the precise value of 32 feet per second per second (Cortina & Dunlap, 1997). Whether or not the social sciences should aim for testing point hypotheses and generating point estimates of effect size as opposed to testing directional hypotheses or a range of acceptable values is a matter of ongoing controversy (Schmidt, 1996; Serlin, 1993).

Cortina and Dunlap (1997) maintain that significance testing can help us adjust our confidence in particular answers. They take the commonsense view that all research starts with a question that is based on theory. Data based on empirical investigation should provide independent corroboration of the theoretically based answer: The stronger the theory base, the less critical the supportive data. Researchers are kept honest by having someone else decide on the appropriate cutoff (e.g., alpha level of .05) or doing so themselves before peeking at the results. To corroborate an answer empirically, we need objectivity, and we need to exclude alternative hypotheses and exclude alternative explanations. Excluding alternative explanations is the most challenging requirement because it involves sampling error as a possible source of covariation that must be discounted. Many of the cited concerns about significance testing, they assert, really concern experimenter error rather than being endemic to the tools of statistical analysis. The use of overly small samples is a case in point. Another is interpreting the conditional probabilities of empirical results as conditional probabilities of hypotheses. The latter point warrants discussion because it has been raised repeatedly as a criticism of NHST (cf. Cohen, 1994; Gigerenzer, 1993; Shaver, 1993).

NHST allows us to draw conclusions about the likelihood of a result given the truth of the null hypothesis. This can be expressed as the conditional probability $P(A|B)$, where A refers to the empirical result and B refers to the null hypothesis. $P(A|B)$ is also the obtained p value of a study. A syllogistic reasoning error occurs, however, whenever you conclude that $P(A|B) = P(B|A)$, the probability of the null hypothesis given the empirical result. This has been referred to as the **inverse probability** error and is regrettably common in the reporting of results from statistical significance testing, primarily because this is the conclusion we hope to draw in the scientific inference process. Without additional information, the exact likelihood of the null hypothesis given the observed data is unknown. This would be analogous to concluding that "If a man deserves to be punished, he is a robber," from the statement, "If a man is a robber, he deserves to be punished." In Aristotelian terms, this is called an invalid form of *modus ponens* (from the Latin *ponere*, meaning to affirm). Cohen (1994) has gone to some lengths to demonstrate how insidiously easy it is to fall into this logical trap because we are motivated to draw conclusions about the population and not the sample. Cortina and Dunlap (1997) retort that these misinterpretations are shortcomings of the interpreter and not of the statistical methods and that, furthermore, there are many contexts in which interpreting a small value of $P(A|B)$ to imply a small value of $P(B|A)$ is practical and accurate. We refer you to their article for a fuller description of their position.

The method for determining the conditional probability of an event is sometimes referred to as Bayes's Theorem. It was developed by the Reverend Thomas Bayes, an 18th-century Presbyterian minister, as an integral part of an alternative theory of statistical inference. Bayesian statistics rely heavily on the formulation that "posterior is proportional to prior times likelihood." This means that statistical inference begins with stating some beliefs about various alternative hypotheses based on knowledge available at that point (called *prior prob-*

abilities, or P(X)) and modifying those beliefs based on collecting relevant data to arrive at so-called *posterior probabilities*, P(B|X). From this perspective, you can take account only of the probability of events that actually have taken place under various hypotheses, and not of events that conceivably could have happened but did not (Lee, 1989). This is the basis of likelihood. The logic of conditional probabilities can be applied to point estimates or interval estimates using distributions such as the normal (Gaussian) distribution specified by most statistical tests. P(I will die within the next 12 months|I have prostate cancer) sets a point estimate with a tail. P(I will die between 6 and 18 months from now|I have prostate cancer) defines an interval estimate.

From the Bayesian perspective, practical significance is much more important than statistical significance. The approach acknowledges the role of human judgment and values in drawing inferences from data. In classical statistics, the alpha level (Type I error) is stipulated at the outset. In Bayesian statistics, both Type I error (false positive rates) and Type II error (false negative rates) can be juggled to reflect the values of the researcher. If it is three times as bad to misclassify a member of one population, represented, let's say, by the null hypothesis, as it is to misclassify a member of another population, represented by the experimental or alternative hypothesis, then that 3:1 ratio is built into interpreting the results. When the odds are larger than 1/3, choose the first population; when the odds are smaller than 1/3, choose the alternative population. For example, assume you are evaluating an expensive assessment battery for diagnosing breast cancer. The relative value of misdiagnosing patients with cancerous tumors versus misdiagnosing patients with benign tumors is likely to be skewed in the direction of reducing the false negative error rate while inevitably increasing the false positive error rate (i.e., it is better to misdiagnose benign tumors as cancerous than to misdiagnose cancerous tumors as benign).

Here is another way to think about interpreting outcomes from a value-driven perspective. A researcher might choose to set alpha = .10. This is the conditional probability P(alternative hypothesis|null hypothesis). The researcher might also set beta = .25. This is the conditional probability P(null hypothesis|alternative hypothesis). The ratio of alpha/beta = .40 reflects the relative seriousness of the two types of errors. The emphasis on practical significance means that the same statistical significance level can lead to different conclusions when the alternatives are different (Schmitt, 1969). What is the relative importance of concluding that a value deviates from 0 when it really is 0 or of overlooking a treatment effect or group difference? In part, this decision depends on previous knowledge about a subject and how secure we are in defining prior probabilities. Suppose that 5% of all tumors are cancerous and 95% are benign. We can express this mathematically using prior probabilities, where $P(A = \text{cancer}) = .05$ and $P(B = \text{benign}) = .95$. Now imagine that we have devised an assessment battery for diagnosing tumors. Let us say that the assessment battery has a false positive rate of .20, expressed as the conditional probability $P(+|B) = .20$, and a false negative rate of .10, expressed as the conditional probability $P(-|A) = .05$. Conversely, this implies that the battery will correctly identify a cancerous tumor 90% of the time $(1.00 - .10 = .90)$ and correctly identify a benign tumor 80% of the time $(1.00 - .20 = .80)$. We are, of course, interested in evaluating the effectiveness of this assessment battery. That suggests that we need to compute the posterior probabilities. Recall that Bayes's Theorem stipulates that

$$P(A_i|X) = P_0(A_i)P(X|A_i)/\Sigma P_0(A_j)P(X|A_j)$$

where the As represent a set of mutually exclusive and exhaustive alternatives, $P_0(A_i)$ is the prior probability of A_i, X is the observation, and $P(X|A_i)$ is the probability of the observation given that A_i is true.

What are the chances of a correct diagnosis of a cancerous tumor if the test result is positive?

$$P(A|+) = P(A)P(+|A)/[P(A)P(+|A) + P(B)P(+|B)] =$$
$$.05 \times .90/ \ [(.05 \times .90) + (.95 \times .20)] \ = .191$$

This means that if the test claims a tumor is cancerous, it will be accurate only 19.1% of the time.

Sometimes, of course, the prior probabilities from the population are not available or must be estimated, but researchers generally know approximately what the data should say to accept the null hypothesis or reject it (Schmitt, 1969). Under the NHST model, the *observation* defines the boundary of the rejection region. The size of the corresponding alpha determines the decision to accept or reject the null hypothesis. In contrast, adherents of Bayesian statistics maintain that the only relevant conclusion to be drawn from an experiment consists of the actual observation made and not the other possible outcomes that *might* have occurred. Thus, Jeffreys (1939), an early proponent of the Bayesian approach, criticizes classical statistical inference theory by dryly noting, "What the use of P (probability of Type I error) implies, therefore, is that a hypothesis that may be true may be rejected because it has not predicted observable results that have not occurred" (Jeffreys, quoted in Lee, 1989, p. 124).

From the Bayesian perspective, you cannot arrive at a true decision criterion without exploring the alternatives in greater detail. A significance level of $p = .023$ does not imply that P(null hypothesis) = .023; the same significance level may suggest different Ps when the alternatives are different (Schmitt, 1969). The onus therefore is on the researcher to decide how much this value has to deviate from zero to have practical significance.

The mathematics underlying the derivation and application of Bayesian statistics can be a bit complex. We present this brief introduction to the subject because it is a legitimate alternative approach that takes issue with the basic NHST model. Most social science researchers are not overly familiar with the approach. For those of you who are interested in pursuing a deeper understanding of Bayesian statistics, a number of texts are available. Schmitt (1969) and Lee (1989) are relatively introductory, whereas Jeffreys (1990) is an excellent advanced text.

Almost everyone agrees that the use of planned comparisons, confidence intervals, graphic descriptions of data, and greater emphasis on Type II error are useful antidotes to some of the more pernicious limitations of statistical decision testing. Other creative recommendations include Harris's (1997) proposal to resurrect Kaiser's (1960) tripartite alternative to dichotomous tests. He proposes three options—strong evidence in one direction, strong evidence in the other direction, and insufficient evidence to conclude there is a difference—as a way of avoiding one-tailed tests, stating significance without a direction, and stating hypotheses in null form. Thus, a significant result such as $p < .001$ does not imply a larger difference than $p = .05$, but it does imply a greater likelihood of being correct about the direction of the population effect. A finding of statistical significance at a smaller probability level, say .001 compared to .05, also implies a greater likelihood of successful replication. Harris argues that a study with a .05 significance level has a 50% chance of being replicated, whereas a study with a p of .005 has an 80% chance of replication.

Eliminating significance tests, as many vocal statisticians are recommending, may not be a straightforward solution either. Take the example of confidence intervals. We strongly recommend that you provide confidence intervals for all important sample statistics. The size of the confidence interval offers a good indication of statistical power. Small confidence intervals imply greater power. But should one also use confidence intervals to make dichotomous decisions about statistical significance? Perhaps, but not if you totally disagree with the logic and utility of NHST and think that such decisions impose a mindless illusion of certainty that is misleading. Cortina and Dunlap (1997) take the implication even further. They claim that the elimination of significance tests also wipes out the meaning of confidence intervals because confidence intervals are dependent on alpha levels. The interval endpoints are themselves random variables estimated from sample data. Setting alpha equal to zero yields a confidence interval that stretches from minus infinity to plus infinity. Likewise, without alpha, power becomes zero. Their solution? Use confidence intervals and power in conjunction with significance testing.

Are you undecided? Do not despair! We believe that the ferment that is currently taking place in the social sciences with regard to questioning historical methods of statistical analysis can only lead to improved ways of interpreting research findings. As Sandra Scarr (1997) has noted, methodology, including measurement and statistics that yield reliable data and probabilistic judgments about evidence, remains our greatest contribution and sets us apart from the naive disregard of the rules of evidence and probability so frequently exhibited by the lay public. Let us remember too that all statistical tests have advantages and disadvantages and that significance testing is only one criterion among many for evaluating a research claim. Regardless of the future of statistical significance testing, the research practitioner does need to be informed about the strengths and limitations of various approaches and have a clear understanding of the meaning of the results he or she obtains.

What Are the Assumptions of Statistical Testing?

How do I . . . ? When should or shouldn't I . . . ?		**What is . . . ? What are . . . ?**	

How do I . . . ? When should or shouldn't I . . . ?

Adjust when assumptions are violated?

Approximate a random sample?

Be concerned about meeting the assumptions of a test?

Be concerned about the randomness of the sample?

Calculate the pseudo-standard deviation?

Calculate an appropriate sample size?

Design or operationalize instruments to maximize information?

Examine data for bivariate normality?

Examine data for equal population variances?

Examine data for homoscedasticity?

Examine data for independence of observations?

Examine data for multivariate normality?

Examine scales for measurement error?

Examine the assumptions of a test?

Examine residual errors?

Know if a distribution is normal?

Match data to statistical tests?

What is . . . ? What are . . . ?

A normal probability plot?

A random sample?

A residual?

Assumptions about error or disturbance terms?

Assumptions about measurement?

Assumptions about the sample?

Assumptions about the statistical model?

Assumptions related to characteristics of population distributions?

Autocorrelation?

Bivariate normality?

Bivariate scatterplot matrix?

Disturbance term?

Durbin-Watson statistic?

Error term?

Homoscedasticity?

Independence of observations?

Intraclass correlation coefficient?

Levels of measurement?

Measurement error?

Multivariate normality?

Nonparametric tests?

Parametric tests?

Population parameters?

Pseudo-standard deviation (PSD)?

Q-Q plot?

Robustness of a test?

Scales of measurement?

Specification error?

The Levene test?

The major assumptions underlying statistical tests?

Level: Advanced

Focus: Conceptual

CHAPTER five

What Are the Assumptions of Statistical Testing?

This chapter addresses the question of what is meant by the "assumptions" of statistical analysis and how we determine whether or not the assumptions of a statistical test have been met. In Chapter 3 of this book, we made the point that even though statistical calculations often require extensive mathematics, the social scientist can view statistics as tools that assist the job of social research, rather than focus on mathematical operations. In this chapter, we return briefly to the issue of the foundations of statistical analysis and hedge somewhat on the above implication that one can ignore mathematics. A computer program will produce an "answer" regardless of the appropriateness of that answer to a given type of data or statistical test. Statistical tests are built on a foundation of assumptions that permit the tests to function correctly, that is, to reach conclusions that are not biased by either over- or underrepresenting the magnitude of relationships or the probability of a given outcome. Sometimes these assumptions are quite extensive, and another test that accomplishes a similar goal may not require such "restrictive" assumptions. If the social researcher is unaware of the nature of the assumptions that underlie a given test, he or she inadvertently may produce a result that is inaccurate or, even worse, nonsensical. A simple example will make this point clear. Table 5.1 shows the distribution of religious

TABLE 5.1 Distribution of Religious Preference Among U.S. Residents

Religious Preference

		Frequency	Percent	Valid Percent	Cumulative Percent
Valid	1 Protestant	953	63.5	63.9	63.9
	2 Catholic	333	22.2	22.3	86.2
	3 Jewish	31	2.1	2.1	88.3
	4 None	140	9.3	9.4	97.7
	5 Other	35	2.3	2.3	100.0
	Total	1492	99.5	100.0	
Missing	8 DK	1	.1		
	9 NA	7	.5		
	Total	8	.5		
Total		1500	100.0		

Statistics

	N		Mean	Std. Deviation
	Valid	Missing		
Religious Preference	1492	8	1.64	1.06

preference for a sample of 1,500 U.S. residents. (The distribution is presented as it was produced by SPSS® for Windows 95®, Version 7.0. Note that printouts from SPSS will differ slightly in their appearance from the reproductions in this book, but the data contained will be identical.)

Note that the table provides the frequency and percentage of responses in each category of religious preference and the "Valid Percentage," which is the percentage calculated after excluding the one DK (Don't Know, coded 8) response and the seven NA (No Answer, coded 9) responses, which were classified as "Missing Values." Also included are the code numbers used to represent the categories of religious preference. For example, Protestants were coded "1" and Catholics were coded "2." Those who didn't know their religious preference were coded "8." These numbers are arbitrary, as is the order of the categories. Thus, Protestants just as easily could have been coded "2" and Catholics "1" without changing the meaning of the distribution. By looking at the distribution, we can see that almost 64% of the distribution stated their preference as "Protestant." This would be true regardless of the arbitrarily chosen value to code Protestants when the data were entered into a data file prior to computer analysis. (See Chapter 1 for an extensive discussion of the logic of variable coding.)

Now examine the two statistics calculated from this distribution, the mean and the standard deviation, shown below the table, and ask yourself, "Does it make any sense to report the mean and standard deviation of religious preference for this distribution?" The answer is that it does not make any sense because "Religious Preference" is a discrete (or nominal) variable. After all, what does an average religious preference of 1.64 mean? Stated more formally, when one calculates the mean value of a nominal variable, such as religious preference, one of the assumptions underlying the mean has been violated, and the answer makes no sense. Means should be calculated only for variables that assume at least the interval level of measurement (i.e., continuously distributed variables). We can speak sensibly of a mean age or a mean income, but not of a mean religious preference. The point is that the researcher should be aware of the underlying assumptions of a statistical procedure before he or she uses that procedure *and* should make some effort to determine if the assumptions have been met. When working with statistics, the consequences of violating assumptions typically are not obvious, and the assumptions themselves may not be clearly understood, but the outcomes of our analyses may be seriously distorted when these assumptions are not taken seriously.

In this chapter, we attempt to accomplish two goals. First, we clarify the nature of the major "assumptions" underlying statistical tests. We discuss only those assumptions that we consider common enough and important enough to deserve comment, but in the short discussion that follows we manage to include most of the major assumptions underlying frequently used statistical techniques. Second, we provide the reader with some guidelines for examining the assumptions of statistical tests. In Part III, we make some recommendations concerning how to proceed with analysis if the assumptions of a test have not been met.

The interested reader who completes this section may conclude that it is never possible to satisfy all the requirements of statistical testing. In this context, it is important to recognize that some problems are more serious than others, and some tests suffer more than others when these problems are encountered. The ability of a test to survive violations of its assumptions is called the

robustness of the test. A **robust** test is one that continues to perform well when its assumptions are violated. By "perform," in this context, we mean that the test maintains the nominal levels of alpha and beta error desired by the researcher.

A test may be robust with regard to some assumptions, but not others, and most tests may be particularly susceptible to bias if certain assumptions, such as independence of observations, are violated. Regarding other assumptions, the exact consequences of violation with certain tests may be unknown. In some cases, the issue may be one in which "minor" violations cause no major problems, but "severe" violations bias the results. Of course, the question of where to draw the line between minor and severe may be in dispute.

Our position is that we are always better off when we know the characteristics of our data vis-à-vis the statistical assumptions that underlie the tests we propose to use. With this knowledge comes the ability to make rational decisions about how to proceed. We may choose to transform the data to better accommodate a particular test, or we may choose to abandon a test in favor of another similar test that does not require an assumption our data cannot meet. In Parts III and IV, we devote extensive discussion to these issues.

In a fairly extensive review of statistics texts, we found very few that contain any general discussion of the assumptions of statistical analysis. Introductory statistical texts typically make brief reference to assumptions, and then proceed, without discussing the assumptions further. Intermediate level statistical texts, covering specific topics, often do a good job of explaining assumptions relevant to the tests in question, but this material is spread throughout these texts. Advanced texts and journal articles often provide the derivation of a statistical test, with mathematical justification for the underlying assumptions. None of these approaches is appropriate for what we are attempting to accomplish in this chapter. This is because the assumptions are not discussed within a general framework for understanding their purpose and because most social scientists lack the mathematical sophistication to understand proofs in statistical journals. We attempt to reach a middle ground that prepares the reader for an informed consideration of the assumptions underlying specific tests when they are encountered in the course of conducting research.

The assumptions of statistical testing can be fairly extensive but appear to fall into one of five major overlapping areas:

1. Assumptions about characteristics of the variables in the population, or the population distribution;
2. Assumptions about error or disturbance terms;

3. Assumptions about the nature of the sample;

4. Assumptions about measurement and measurement error; and

5. Assumptions about the model being tested.

As will become apparent in the following discussion, there is some overlap among the above categories, but they meet our needs and assist with understanding the material. In most of what follows, the assumptions we discuss are required to properly conduct tests of statistical hypotheses, such as t tests, analysis of variance, and regression analysis; however, some procedures contain assumptions that, when violated, might bias a test of significance but would not invalidate a measure of association or another descriptive statistic. We make this distinction clear when it is important and not obvious from the nature of the discussion. We first consider assumptions in each of the above five areas, then illustrate the major assumptions with an example, and finally offer some suggestions regarding how to test for or detect violations of assumptions. We discuss the question of what to do when specific assumptions are violated in Part III.

WHAT ARE THE ASSUMPTIONS RELATED TO CHARACTERISTICS OF POPULATION DISTRIBUTIONS?

This section examines assumptions regarding the characteristics of the population from which a sample is being selected. More formally, such assumptions are referred to as assumptions regarding population **parameters**. According to Siegel (1956), "In the development of modern statistical methods, the first techniques of inference that appeared were those requiring a good many assumptions about the nature of the population from which the scores were drawn" (p. 2). This statement is still true today, and many, if not most, of the tests one frequently finds in use require that data being analyzed are sampled from a population with certain characteristics, most typically that the variables in question are normally distributed, and that they are at least intervally scaled. Because characteristics of populations are called "parameters," these tests have become known as **parametric tests**. When these assumptions are not met, social scientists often turn to **nonparametric** or "distribution-free" tests because, as the terms imply, these tests do not require normally distributed population parameters and/or intervally scaled variables. (We consider nonparametric tests in more detail in Chapter 8.)

? What Are Normality and Multivariate Normality?

A normal distribution, represented by a bell shaped curve, is a symmetrical, unimodal distribution. Being bell shaped, however, does not in and of itself mean that a distribution is normally distributed. In the language of probability theory, distributions of continuous variables are described by equations that, when graphed, produce smooth curves. The areas under these curves represent probabilities, with the total area under the curve being equal to 1.00 (because a probability cannot exceed 1.00 and all probabilities must sum to 1.00). The equation representing the normal distribution (also called a probability density function) when the variable X has a mean (μ) and standard deviation (σ) is given by the following probability density function:

$$f(X) = \frac{1}{\sigma_x \sqrt{2\,\pi}}\, e^{-(x-\mu)^2/(2\sigma^2 x)}$$

In the above equation, pi (π) and e are mathematical constants; thus, the normal distribution is defined by the values of the population mean and standard deviation, μ and σ, respectively. When graphing $f(X)$ against X, one produces the familiar bell shape known as the normal curve.

One of the implications of the above excursion into mathematical theory is that there is a different normal curve for every combination of mean and standard deviation. In these curves, the mean defines where the center of the distribution is and the standard deviation defines its spread (or variability or dispersion). The **standard normal distribution** tables found at the back of virtually all statistics texts represent the Z score transformation applied to a normally distributed variable, thus creating the familiar bell shaped curve with a mean of 0 and a standard deviation of 1.

Given this introduction to the normal curve, let's extend our discussion to one of the most frequently encountered statistical tests, the t test for the difference between means. The assumptions of this test include (but are not limited to) the assumption that the variable of interest is sampled from a normally distributed population. In practice, the normality assumption can be relaxed when N is more than 50; however, the t test generally is considered a "small sample test," commonly used with sample sizes less than 50. If we relax this assumption slightly, so that we assume only that the variable's population distribution is *approximately* normal, then the question to ask is, "OK, how do I know whether or not this assumption is justified?" The unfortunate answer is, "Unless you know the population distributions, you cannot completely verify this assumption." (If you knew the population distributions, however, there would be no need for a statistical test!) To deal with this dilemma, researchers can examine the data that they do have, and the theory and research of others, to make reasonable assumptions about the nature of population distributions. In Chapter 2, we introduced techniques of exploratory data analysis (or EDA), based on sample information, that utilize graphical and numerical techniques to explore the plausibility of the normality assumption and other assumptions to be considered later. Below, we expand on this introduction to consider the question of normality.

? Is the Distribution Normal?

A critical question that is raised throughout this book and that forms the core assumption of all parametric statistics concerns the "normality" of the distribution of variables. We address issues related to the normality of variables and the implications of lack of normality for statistical testing in Chapter 8. In this section, we are concerned only with the question of "How do we know?"

A plot or graph provides a preliminary picture of a distribution and offers our first insight into whether or not the distribution might be considered normal or close to normal. To facilitate this discussion, we return to the distribution of age at marriage first presented in Chapter 2. We present this distribution again in Figure 5.1.

We know two very important things about the distribution of age at marriage: it appears positively skewed, and it appears leptokurtic, and thus age at first marriage probably is not normally distributed. The sample size would be a consideration if we were considering the normality of the population from which the sample came; however, this is an extremely large sample ($N = 1,202$) and probably represents the population distribution quite well.

In addition to the graphs we presented in Chapter 2 and Figure 5.1, a number of other numerical and graphic techniques specifically address the question of normality. The first and one of the most simple is based on our knowledge that the mean, median, and mode of a normal distribution are all the same. For the age at marriage distribution, these figures are: mean = 22.79, median = 22, and mode = 21.

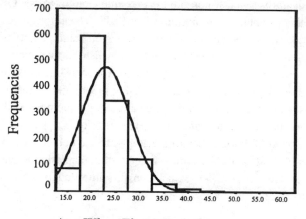

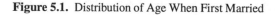

Age When First Married

Figure 5.1. Distribution of Age When First Married

These values are quite similar, but, as we would expect from a positively skewed distribution, the mean is higher than the median, and the median is higher than the mode. Because these values are within 2 years of one another, however, we might be tempted to conclude that the distribution of age at marriage should be considered approximately normal. Note, however, that the comparison of mean, median, and mode utilizes only measures of central tendency and does not consider measures of variability or spread around these central values. Hamilton (1990) suggests that when distributions appear approximately symmetrical and unimodal, the pseudo-standard deviation (PSD) can be used to examine deviations from normality. Recall that a normal distribution is one form of a symmetrical/unimodal distribution and comparison of the PSD with the standard deviation assists in detecting the deviation of these distributions from normality. If the standard deviation and the pseudo-standard deviation are similar, one can conclude that the distribution is approximately normal. When the standard deviation is greater than the PSD, however, the distribution tends to have heavy tails or may produce a histogram that appears flat, or platykurtic. When the standard deviation is less than the PSD, the distribution tends to have light tails or may produce a histogram that appears peaked, or leptokurtic. For the above distribution of age at marriage, the standard deviation is 5.03, and the PSD is calculated by dividing the interquartile range (25 − 19 = 6) by 1.35 (6/1.35 = 4.44). Because the standard deviation (5.03) is greater than the PSD (4.44), we know that the distribution deviates from a normal deviation and that

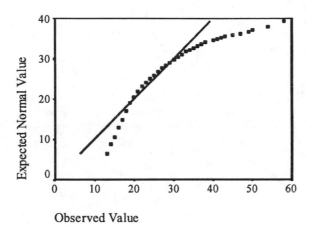

Figure 5.2. Normal Probability Plot (Q-Q Plot) of Age When First Married

this deviation is reflected in heavier tails, specifically in this case by the heavy tail in lower end of the distribution.

We can also calculate statistics that show us the extent of the skew and kurtosis in this distribution. The value of skew of the distribution of age at marriage is 1.66, with a value of 0 showing no skew and negative values showing negative skew. The value of kurtosis is 5.382, with a value of 0 showing no kurtosis, a negative value showing platykurtosis, and a positive value showing leptokurtosis. Thus, this distribution is highly leptokurtic, as shown by Figure 5.1.

In addition to these numerical procedures, we can plot a comparison of the age at marriage distribution with what we would expect if the distribution was perfectly normal. These plots are called **normal probability plots** or Q-Q plots. The horizontal axis shows the location of the points as observed in the distribution. The vertical axis shows the location of the points as expected if the distribution were normal. If the observed and expected distributions are the same (i.e., the distribution is perfectly normal), a diagonal straight line, from left to right, would be the result. This line is also drawn on the graph. We present the normal probability plot for the age at marriage distribution in Figure 5.2.

Figure 5.2 shows positive skew, with outliers in the upper tail of the distribution. Figure 5.3 presents some other possible outcomes of normal probability plots for future reference, and Figure 5.4 shows the normal probability plot for the age at marriage variable after excluding the 38 outliers from

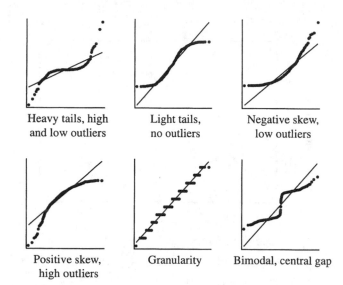

Heavy tails, high   Light tails,   Negative skew,
and low outliers   no outliers   low outliers

Positive skew,   Granularity   Bimodal, central gap
high outliers

Figure 5.3. Possible Outcomes of Normal Probability Plots

the distribution. In Chapters 6 and 8, we discuss in detail procedures for dealing with both outliers and non-normal distributions.

There are numerous other methods for examining the normality of a distribution. We have described the basics, available with most computer programs and used throughout the remainder of this book. In the next section, we consider the examination of two variables at a time, or bivariate analysis.

? How Do I Examine Data for Bivariate and Multivariate Normality?

Let's assume that you have examined the distributions of your variables for normality and are reasonably satisfied that this assumption has been met. You then decide to conduct a multivariate test, such as multiple regression, and discover that one of the assumptions of this test is that the variables have population distributions that are **multivariate normal**. What does multivariate normality mean? Can you assume that because all the variables individually involved are normally distributed, the assumption of multivariate normality is also valid? The answer to the latter question is that individual normality does not ensure multivariate normality but is a precondition for it. To get a better handle on this concept, which is a difficult one, let's begin with the concept of

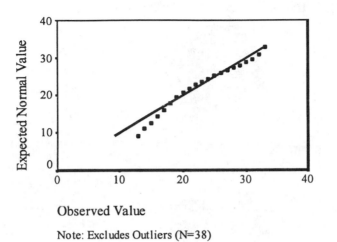

Observed Value

Note: Excludes Outliers (N=38)

Figure 5.4. Normal Probability Plot of Age at First Marriage, Excluding Outliers

bivariate normality, one of the basic assumptions underlying a correlation coefficient, and extend our discussion from there to multivariate normality.

In a **bivariate normal distribution**, for any given value of X, the Y variable will be normally distributed, and for any given value of Y, the X variable will be normally distributed. When considering a three-dimensional plot, the concept of bivariate normality can be visualized as a "hat" or "mound." Any cross section of this hat (parallel to either the X or Y axis) will, in theory, be normally distributed, and any horizontal cross section should appear as an ellipse, such that the higher the correlation between variables, the thinner the ellipse. Figure 5.5 shows two bivariate normal distributions, one representing little or no correlation between the two variables, and one representing a correlation of .75. Note that both represent a hat, and that any *vertical* cross section parallel to either X or Y would produce a bell shaped curve. A *horizontal* cross section of the first distribution, however, would produce a more circular ellipse, whereas a similar cross section of the second distribution would produce a much thinner, cigar-shaped ellipse. These cross sections are shown to the right of each distribution.

Moving to the concept of multivariate normality,

A set of variables is said to have a multivariate normal distribution if each of the variables has a linear regression on each of the others (simple regression) and on every possible subset of the other variables (multiple regressions), and where the

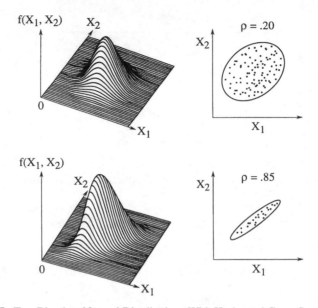

Figure 5.5. Two Bivariate Normal Distributions With Horizontal Cross Sections

deviations from the regression equations are normally distributed. (Kachigan, 1986, p. 330)

In addition, any linear combination of the variables will be normally distributed, and all subsets of the set of variables will have a multivariate normal distribution. Why spend so much time on the assumption of bivariate and multivariate normality? Quite simply, this is one of the central underlying assumptions regarding the population distributions of all parametric multivariate techniques!

Unfortunately, we cannot point to a simple test of bivariate or multivariate normality and must address this concept through the use of other methods. First, we suggest the construction of the bivariate scatterplots for all variables. As Stevens (1992, p. 245) suggests, bivariate normality, for correlated variables, implies that the scatterplots for each pair of variables will be elliptical; the higher the correlation, the thinner the ellipse. Thus, as a partial check on multivariate normality, one could obtain the scatterplots for pairs of variables and see if they are approximately elliptical. Fortunately, bivariate scatterplots are quite easy to produce, and all versions of SPSS®, Stata®, SysStat®, SAS®, and BioMed® produce matrices of such scatterplots. For example, in Figure 5.6 we present a 3×3 correlation matrix of measures of peer relationships, self-esteem, and risk for drug and alcohol abuse from a sample of middle school students.

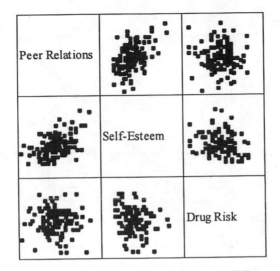

Figure 5.6. Bivariate Scatterplot Matrix of Peer Relations, Self-Esteem, and Drug Risk

The correlation matrix and scatterplot matrix in Figure 5.6 correspond to one another. Clearly, self-esteem and risk for alcohol or drug use are negatively related, but this is most obvious from the middle right-hand cell. The remaining two correlations both show positive relationships, with the relationship between peer relations and self-esteem being stronger than the relation between peer relationships and risk for drug use. If the researcher is satisfied that all the displayed relationships as shown in bivariate scatterplot matrices represent roughly linear relationships, then one piece of evidence supporting an argument for multivariate normality would be provided.

A second similar strategy that supports the assumption of multivariate normality is to plot the residuals of each dependent variable. In a regression equation, this may mean examining only one variable, but with procedures containing multiple dependent variables, more than one plot may be involved. For example, the graph in Figure 5.7 plots the residuals from the bivariate regression when self-esteem is used as an independent variable and risk for alcohol or drug use is used as a dependent variable. The plot shows the distribution of residuals compared to the expected distribution under the assumption of normality. As can be seen from this plot, the residuals fall almost perfectly on the diagonal line, in this case supporting our assumption of bivariate normality.

We have presented what we believe to be reasonable and easy ways to conduct tests, but we have presented only some of the possibilities. For the case

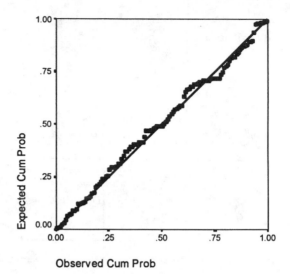

Observed Cum Prob

Figure 5.7. Normal Probability Plot of Residuals for Risk for Drug Abuse, by Self-Esteem

of normality, other tests and other graphic procedures are readily available. In the context of regression and multiple regression, a number of excellent texts explore the meaning of regression output and examine the violation of the assumption using graphical presentation (Cook & Weisberg, 1994; Fox, 1991; Hamilton, 1992). In the context of examining multivariate normality, there are more complex tests available (see Gnanedesikan, 1977, pp. 168-175; Stevens, 1992, pp. 247-251); however, these are difficult to implement and are seldom used. As Kachigan (1986) states, "The violation of the assumption [of multivariate normality] has never stopped anyone from gathering and analyzing data in which they are interested. Techniques for identifying departures from the multivariate normal distribution are still a subject of early developmental work" (p. 330).

❓ What Is Homoscedasticity?

The assumption of equal population variances is central to both ANOVA (analysis of variance) and regression statistics, and is often referred to as the "homogeneity of variance" assumption or **homoscedasticity**. In t tests and one-way ANOVA, this assumption is expressed as

TABLE 5.2 Distribution and t Test of Occupational Prestige, by Smoking Behavior

Variable	Number of Cases	Mean	SD	SE of Mean
Nonsmoker	211	42.190	14.116	.972
Smoker	340	41.577	14.733	.799
Mean Difference = .6131				

Levene's Test for Equality of Variances: F = .465; P = .496
t Test for Equality of Means

Variance	t Value	df	2-Tailed Significance	SE of Difference	95% Confidence Interval for Difference
Equal	.48	549	.630	1.271	(–1.883, 3.109)
Unequal	.49	459.75	.626	1.253	(–1.859, 3.035)

$$\sigma_1^{\,2} = \sigma_2^{\,2} \ \text{ or } \ \sigma_1^{\,2} = \sigma_2^{\,2} = \sigma_3^{\,2} = \sigma_n^{\,2}$$

and states that the variances of the same variable, selected from independent samples, will be equal. In regression analysis, this assumption states that the variances of the Ys, for each X, will be equal. Neither analysis of variance nor regression analysis tests this assumption directly; however, procedures for testing this assumption with sample data are fairly straightforward and included with most statistical packages. This assumption is needed in order to form independent estimates of explained and unexplained variance, necessary to conduct the F test.

The assumption of equal population variances is, compared to tests of multivariate normality, easy to implement and common output in most statistical packages. Probably the most easily computed of these tests is the Levene test for equality of variances. This test is based on the fact that the ratio of the larger to the smaller of two variances is distributed as F, with $n_1 - 1$ and $n_2 - 1$ degrees of freedom. This test is included with the t test in SPSS and other statistical program packages. For example, the t test in Table 5.2 (adapted from SPSS output) shows the relationship between smoking behavior and self-rated satisfaction with health.

Note that in addition to showing the descriptive statistics for the smoker and nonsmoker groups, Levene's test is also provided. This test then assists the researcher in selecting one of the two rows, labeled "Equal" and "Unequal," as

TABLE 5.3 One-Way Analysis of Variance of Occupational Prestige, by Social Class

Source	df	Sum of Squares	Mean Squares	F Ratio	F Probability
Between groups	2	24,419	12,209	66.55	.0000
Within groups	1,252	229,707	183		
Total	1,254	254,127			

Group	Count	Mean	Standard Deviation	Standard Error	95% Confidence Interval for Mean
Group 1	50	32	11.6	1.6440	29 to 35
Group 2	597	36	12.1	.4969	35 to 37
Group 3	608	44	14.9	.6053	43 to 46
Total	1,255	40	14.2	.4016	39 to 41

Levene's Test for Homogeneity of Variances

Statistic	df1	df2	2-Tailed Significance
10.4976	2	1,252	.000

NOTE: Prestige = respondent's occupational prestige score; Class = subjective class identification. Most numbers to the right of the decimal point have been eliminated for ease of reading this example.

appropriately meeting the assumptions of the test. If the Levene's test is significant, the null hypothesis of equal population variances is rejected, and the second row of the table is appropriate. If the variances can be assumed to be equal, which is the case in this example, then the first row of the table is appropriate.

In a second example (see Table 5.3), we examine the relationship between occupational prestige and social class with a one-way analysis of variance. Note that there are very large differences in the mean prestige levels that are statistically significant. Note also that the Levene's test at the bottom of the printout is statistically significant. This is because variability in occupational prestige increases as social class increases. The researcher is now left with the question of what to do about this apparent violation of assumptions. Perhaps the differences in the variances are not really that great, but the large sample size ($N = 1,255$) produced a significant difference anyway. Perhaps we would be justified in ignoring the violation of assumptions because one-way ANOVA is "robust" with respect to this type of violation. Perhaps we should use a nonparametric

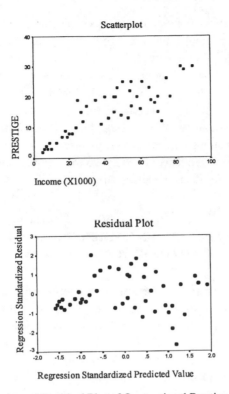

Figure 5.8. Scatterplot and Residual Plot of Occupational Prestige, by Income

alternative to the test. Some statistical software programs, such as the Biomedical package (BioMed), will automatically conduct another test that does not require this assumption. The choice among the numerous alternative strategies depends on the exact nature of the violation and the researcher's own beliefs and preferences regarding both the necessity and the acceptability of the different options.

The standard suggestion for examining the assumption of homoscedasticity in regression analysis is to plot the predicted Y values against the residual values. **Heteroscedasticity** is indicated when these values spread or fan out from left to right or right to left. As an example, let's examine the relationship between income and occupational prestige. The correlation between these variables is .86, and their scatterplot is shown in Figure 5.8. Also shown is the plot of standardized predicted prestige values (based on income as the independent variable) against the standardized residual values. We have deliberately constructed these values to illustrate a point. We distributed prestige so that as income increases, so does the variability in occupational prestige, up to the point where high incomes predict high prestige very well. In other words, we deliberately created heteroscedasticity in the middle of this small data set.

In the example shown in Figure 5.8, examination of either the scatterplot or the residual plot would lead to the conclusion that assumptions had been violated. Sometimes the exact nature of these violations may be difficult to determine because small sample size, lack of independence in observations, and non-normal distributions may all exhibit themselves in plots of residuals. We refer the reader to the references cited above for a more complete treatment of the examination of residuals in regression analysis.

WHAT ARE THE ASSUMPTIONS ABOUT ERROR OR DISTURBANCE TERMS?

Assumptions about error terms or disturbance terms surface in textbooks that discuss regression analysis or other techniques based on least squares estimators, as these assumptions are required to produce unbiased and efficient estimates of the parameters of this model. Therefore, we will discuss the meaning of "disturbance term" in the context of bivariate regression.

The bivariate population regression equation contains two constants, one for the slope and one for the intercept, called alpha (α) and beta (β), respectively. (Alpha should not be confused with the level of significance.) Unless the correlation is 1.00, which is highly unlikely, there will be scatter about the regression line. This means that the value of the dependent variable, Y, for each case, will not fall on the regression line. Thus, when we write the equation that represents the observed (not the predicted) value of Y for each individual, we must include a term that represents the amount that this individual value deviates from the regression line. Expressed as an equation, the value of Y for the ith case, or Y_i, is equal to

$$Y_i = \alpha + \beta X_i + \varepsilon_i.$$

The value of ε_i represents the amount that the observed value of Y for the one case we have designated as "i" from the population of all observations differs from the value predicted by the regression equation. This term is called an **error term**, a **disturbance term**, or a **residual** and is the term about which we are concerned. We may think about this term as representing all the other causes of

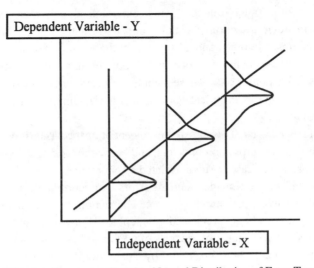

Figure 5.9. Bivariate Regression Showing Normal Distribution of Error Terms Around Regression Line

Y that have not been specifically brought into the regression equation (i.e., not specified by our model).

To utilize ordinary least squares (OLS) to estimate the parameters in the population regression equation with sample data in a manner that provides what is known as the "Best Linear Unbiased Estimate" (BLUE), we must make several assumptions about the error term. First, we must assume that the expected value (i.e., the mean) of this term (ε_i) is zero and that it is normally distributed. Second, we must assume that the errors are independent of one another (no **autocorrelation**) and have a constant variance (no **heteroscedasticity**). Third, we must assume that *X* (the independent variable) and ε_i are uncorrelated. Violation of the above assumptions may result in OLS estimates that are biased or significance tests that are inaccurate. Figure 5.9 shows a graph of a bivariate regression in which the errors are normally, independently, and identically distributed at every level of *X*. We generally believe that errors should be normally distributed if they are the result of many small and random influences, and if the measurement of one case does not influence the measurement of a second case (the independence of observations assumption, discussed below).

A detailed discussion of the examination of data for violations of these assumptions is beyond the scope of this presentation. We refer the reader to Chapter 4 of Laurence Hamilton's excellent work *Regression With Graphics*

(1992). Hamilton points out that some regression assumptions, as we have illustrated in previous sections, can be examined with sample data. These include linearity, homoscedasticity, autocorrelation, and normality. Some assumptions, however—particularly those involving assumptions about error terms—cannot be checked with sample data. These include the absence of correlation between error terms and the independent variable and the assumption that error terms have a zero mean.

Thus far, we have not discussed the concept of **autocorrelation**. Autocorrelation refers to the correlation of errors, and hence the lack of independence of observations, across cases. Autocorrelated errors reduce the efficiency of OLS estimators and bias estimated standard errors. Autocorrelation is particularly likely to occur when the cases have some connection, such as being from the same family unit or being connected geographically, or when cases are part of a time series. As with many of the tests of population assumptions, we test the assumption of uncorrelated errors (ε_i) by examining the sample residuals (e_i). One such test involves the **Durbin-Watson** statistic, which examines the correlation of successive residuals. A table is available that provides critical values for the sampling distribution of d, the test statistic for this test. Other tests include autocorrelation coefficients that examine the correlations in time-ordered data between residuals from the previous observations.

WHAT ARE THE ASSUMPTIONS ABOUT THE SAMPLE?

Assumptions about the nature of the sample concern how the sample is selected and the size of the sample. We already have discussed the relationship of sample size to the size of the confidence interval, standard error, and power, and we have discussed the importance of the simple random sample to statistical inference. In this section, we will elaborate on these issues as they relate to the question of assumptions underlying statistical testing.

? What Is a Random Sample?

A **simple random sample** is one drawn in such a way that all members of the population have an equally likely chance of being selected or, more precisely, a sample drawn in a manner such that all samples of the same size are equally likely. The method in which such a sample is usually illustrated is to imagine an urn containing all elements of the population, then selecting one case from this urn, replacing it, remixing, and repeating the process until the desired sample size has been obtained. The necessity of a random sample in statistical inference is that, without a random sample, we cannot apply the rules of probability to statistical inference. Said another way, even if all the assumptions regarding population parameters are perfectly valid, we cannot generalize sample results to this population unless we have drawn the sample using a random process. The truth of the matter is that virtually all samples used in social and behavioral research are not simple random samples. The assumption of random selection is almost always violated, and in contrast to assumptions regarding characteristics of the population, these violations are clearly evident and acknowledged. It is common practice to read a journal article in which the researcher(s) discuss having drawn their sample from among volunteers from introductory psychology (or sociology, or some other field) courses and then proceed with statistical inference tests.

Beyond this, such studies typically separate these students into gender groups or ethnic groups and proceed to test these group differences "as if" these groups were randomly drawn. To add insult to injury, it is not uncommon to lump together Chinese, Japanese, Filipino, Korean, and all other similar groups into a category known as "Asian and Pacific Islanders," as if any statement about the characteristics of this group vis-à-vis any other had meaning, independent of the implications of attempting to generalize these results through the conduct of statistical tests.

Nevertheless, a wide variety of sampling designs can approximate simple random samples and permit generalization to the population. In modern sampling theory, the basic distinction is not between simple random samples and all others, but between probability and nonprobability sampling. A simple random sample is the most basic of all probability samples and serves as the foundation for more complex probability sampling designs. Some examples of these designs include systematic random sampling, stratified sampling, and cluster sampling. It is not uncommon for national surveys to employ a combination of probability sampling methods to sample large geographic areas using what is called multistage sampling. It is not our purpose to review these techniques here. Most introductory research methods texts include a chapter on sampling, and the classic work by Kish (1965) has been reprinted recently as one of the Wiley Classics Library Series (1995). Rather, we wish to make only a few points and recommendations to guide the researcher.

First, if the sample is not a simple random sample but *is* a probability sample, statistical inference is still reasonable and meaningful. Formulas exist to adjust standard errors for probability sampling variations from simple random, but in practice these formulas are used only by the most sophisticated of sampling experts, and the failure to use these formulas is not considered an issue. Second, it is extremely common to find use of statistical inference with sampling designs that clearly represent nonprobability samples, such as convenience, purposive, and snowball samples. In many of these cases, inferential techniques are used because the authors and reviewers of manuscripts (and dissertation committees) believe that the findings are not "significant" unless a hypothesis test is involved. In other cases, the argument is made that though one may not have met the requirements of a probability sample, the findings are generalizable to whatever population is represented by the sample in question, and then an attempt is made to define some of the characteristics of that population. This situation is represented by much dissertation research and a sizable proportion of published empirical findings.

So what is the researcher to do? Without statistical tests to illustrate the significance of the findings, the research is likely to be looked upon as "half complete" or not important (i.e., not significant), but if one performs these tests on data from nonprobability samples, the results are also in question! One of the interesting paradoxes is that there are thousands of databases with very carefully drawn samples that go virtually untouched by student and professional researchers alike. Many of these are archived in what is known as the Interuniversity Consortium for Political and Social Research (ICPSR). Included among the archives in the ICPSR are all the National Opinion Research Center's General Social Surveys, extensive data from the Gallup and Field Polls, and hundreds of databases from federally funded studies encompassing a vast realm of political and social data. Many of these have been carefully analyzed to answer only a few small questions, then housed away to gather dust and become obsolete. Our recommendation is to consider gaining access to these data files not only as legitimate research but also as one way, perhaps the only way, to gain access to truly high-quality data collected using probability sampling methods. It should also be noted that most major universities are members of the ICPSR. What this means is that the data are virtually free to anyone from one of these institutions willing to take the time to search for the data that fit their needs. A number of references on the use of secondary analysis have become available (Kiecolt & Nathan, 1985), and we recommend that these be examined by anyone considering the use of secondary source data.

Random sampling does not need to be considered the only game in town. In addition to the vast literature on "naturalistic" inquiry, which eschews the whole logic of inference, we believe that there is much to be gained from smaller case studies of "one organization" or "one group" or a small collection of anything that might provide insights into human behavior. Studies of these groups, with the use of descriptive statistics, seem both reasonable and worthwhile.

? How Do I Calculate an Appropriate Sample Size?

One of the most exasperating questions for the researcher concerns the size of the sample. We addressed this question earlier, because sample size is a critical determinant of the standard error and power, but we have not addressed this question in the context of sample size assumptions attached to specific statistical tests. A couple of important issues should be considered in this context. First, one question a researcher might ask when considering the use of a statistical test is, "What are the consequences of the use of this test with the sample size that I can reasonably be expected to obtain (or with a sample of N cases)?" For example, a perusal of introductory statistics texts will reveal that many have sections with titles like "Large Sample Tests for Means" and "Small Sample Tests for Means." The distinction being made (in introductory texts) is usually between the Z test and t test, because with small sample sizes the standard normal distribution is an inappropriate test statistic. In this context, the issue is clear: If the sample is small, use a t test. Second, many statistical tests simply do not work well with small samples, and their assumptions become tenable only when the sample is of sufficient size. This is particularly true of multivariate methods that utilize more than a few variables, primarily because the statistical models are developed using assumptions that hold true only when the sample is both simple random and large, and the distributions are multivariate normal. For example, a factor analysis that examines the structure of a 30-item questionnaire would begin analysis with 435 unique correlations (i.e., [30 × 29]/2) and an unknown number of factors (≤ 30). It is easy to see that in multivariate analysis, situations quickly arise in which the number of parameters being estimated exceeds the number of subjects in the sample, a situation that may invalidate these tests. To prevent these mistakes from being made inadvertently by novice researchers, a number of guidelines or rules of thumb have been developed. For example, with factor analysis, some authors suggest a 4 to 1 ratio of subjects to variables. More recent suggestions encourage 10 or more subjects for every variable. Thus, in our hypothetical example above, with 30 items (i.e., variables), somewhere between 120 and 300 subjects might be recommended. Of course, this raises some interesting questions, including the following: What is a "small sample"?, What does "sufficient size" mean?, and What if I want to do a factor analysis and only have three subjects per variable? In the discussion that follows, we consider these issues further and make some recommendations.

? A Reconsideration of the Sample Size Issue

A realistic approach to the consideration of sample size incorporates both the costs of collecting data, including the costs of the researcher's own time, and the requirements of statistical testing. Although it might be easy to generate a number that represents the "ideal" sample size, if this number is generated in the absence of some idea of "cost per case," one quickly runs into deficit spending. For example, consider a researcher who decides to conduct a study using a mail survey. The actual research topic is unimportant, but let us also assume that the researcher will conduct a multiple regression analysis using five independent variables and has a $1,000 grant to collect the data. How many subjects would be needed, and can the researcher afford a sample of this size?

Let's assume that the researcher begins with the assumption that a sample size of about 150 subjects is desirable. The typical response rate for a mail survey is between 25% and 40%. If we assume that this survey will generate about a 33% response rate, then to obtain 150 returned surveys, about three times that many will need to be mailed, or 450. If we first estimate the cost of producing one packet—containing the mailing envelope, a cover letter imploring the recipient to respond, and the questionnaire—to be $1, and then add $2 more as the cost of mailing ($1) and enclosing a self-addressed, postage-paid, return envelope ($1), the total cost for 450 surveys will be $3 × 450 = $1,350. Thus, a $1,000 grant is not enough. If we make one minor modification, and use a business reply envelope that incurs a cost only if the respondent returns the envelope, we reduce the cost by $300 (because we would not need to pay postage for the 300 surveys not returned) and are only $50 over budget. Now let us assume that the researcher takes the extra $50 from his own funds and conducts the study. Of the 450 mailed, and 150 returned, some respondents are ineligible and some incorrectly fill out the questionnaire. If this is 10% of those returned, or 15 surveys, will the remaining 135 be enough?

To answer this question, we conducted a power analysis, assuming a multiple regression analysis with five independent variables, an alpha level of .05, and a power level of .8. What would the minimum detectable effect size be with these requirements, and how much could we reduce the sample size and still be satisfied? Our calculations show that with a sample of 135, an R^2 value (effect size) of .10 could be detected with a power of .85. Thus, the researcher can be confident that his or her study is sufficiently well powered with 135 valid surveys. Our calculations also show that we could reduce the sample size to 120 and still obtain a power level of .8, but reducing the sample size further would require us to accept a power level below .8, or accept the possibility that effect sizes of .10 and below may remain undetected. Thus, if the power analysis had been conducted in advance, and we wished to remain within budget, one initial recommendation might have been to conduct the survey with an initial mailing of approximately 400, under the expectation that the valid questionnaires would be $400 \times .33 \times .9 = 120$, at a cost of roughly $(280 \times \$2) + (120 \times \$3) = \$920$. This leaves \$80 left to pay a graduate student to enter the data, probably not enough, but then one can always enter one's own data or beg a spouse to help.

In sum, we can begin our consideration of sample size by asking questions about the type of analyses that we will be conducting, the size of the effects that are meaningful to detect, and the power with which we wish to detect these effects; however, these discussions must always be tempered with a good dose of reality. We recommend that the reader obtain *How Many Subjects?* by Helen Churma Kraemer and Sue Thiemann (1987) for further considerations of these issues in a practical context.

? What Is Independence of Observations?

Most inferential statistical techniques assume that the elements of the population are randomly and **independently** drawn. What this means is that the fact that one element became a member of the sample should not have any relationship to the probability of another element of the population becoming a member of the sample.

A couple of examples will illustrate how lack of independence might occur. In one of the authors' studies of family reunification following removal for child abuse, it was discovered that more than 30% of the sample members were related to one another. This occurred because the removal of a child from the home for abuse also results in the removal of all other children, for what are called "protective reasons." Thus, we were faced with a sample that seriously violated the independence of observations assumption, as any child targeted for child abuse was likely to be accompanied by all of his or her siblings. As a second example, consider the case of a college student who participates in an experiment as part of a class requirement. He then tells his roommate, "You should really participate in this experiment; it only lasts 20 minutes and you get to watch movies depicting violence on TV." Based on this information, the roommate signs up to participate, and even though he might end up in the control group that watches reruns of *Father Knows Best*, his responses clearly are not those of a naive experimental subject.

Why is the independence of observations assumption so important, over and above the obvious complications illustrated in the above two examples? One answer is that the rules of probability required to make valid population inferences are invalidated by lack of independence of observations. The consequences of this are that both the level of significance and power of the F test will be seriously compromised. Lack of independence of observations is assessed by the **intraclass correlation coefficient**. As an example, consider a one-way ANOVA with three groups of 30 subjects each, and an intraclass correlation of only .1. If the alpha level of this analysis were set at .05, the actual level would be .4917. Thus, the researcher presents the results as testing the null hypothesis at .05, when he or she, because the independence of observations assumption has been violated, actually tests this assumption at 10 times that level (Scariano & Davenport, 1987). These observations led Stevens (1992) to state that the independence of observations assumption "is by far the most important assumption, for even a small violation of it produces a substantial effect on both the level of significance and the power of the F statistic" (p. 239).

WHAT ARE THE ASSUMPTIONS
ABOUT MEASUREMENT?

Recall the example used to open this chapter. We attempted to calculate a statistic that was inappropriate for the level of measurement of the variable. The notion of "levels of measurement" or "scale types" is attributed to psychologist S. S. Stevens. According to Stevens, "A rule for the assignment of numerals (numbers) to aspects of objects or events creates a scale [of measurement]" (1966, p. 22). An easier way to say this is that measurement involves the assignment of numbers to objects or events according to rules. Stevens discusses the four types of measurement scales, which themselves are logically ordered according to the types of permissible mathematical (and thus statistical) operations that they permit. These are nominal scales, ordinal scales, interval scales, and ratio scales. Table 5.4 summarizes the properties of these scale types and provides examples. In Chapter 9, we discuss the treatment of data that do not conform to interval or ratio scales.

In sum, nominal measures require only the ability to categorize into mutually exclusive and exhaustive categories. If we add to this the ability to rank order the categories, we then have an ordinal scale. If we add to these qualities the

TABLE 5.4 Scales of Measurement

Scale Type	Basic Empirical Operation	Operations We Perform	Examples
Nominal	Determination of equality	Identify and classify	Religious preference, political preference, team jersey numbers
Ordinal	Determination of greater or less	Rank order	Qualities of objects, such as lumber, leather, or gems; personal preferences and attitudes
Interval	Determination of equality of intervals or differences	Find distances or differences	Temperature, calendar dates
Ratio	Determination of equality of ratios	Find ratios, fractions, or multiples	Length, weight, loudness, brightness, duration, most physical scales

ability to determine the exact distance between categories that we can express as some standard interval, we then have an interval scale. Finally, if we meet all the requirements above, and the measures are based on a true or absolute zero point, we have a ratio scale. This basic framework, as is familiar to almost anyone who has taken a social sciences statistics course, has served as a fundamental model for the selection of statistical techniques and as a framework for the organization of statistics texts themselves.

The Stevens classification is not without its critics, some of them quite severe, and we will take up some of these criticisms in a later section. The great value of this framework is that it provides a convenient device for thinking about our data and what we can, and cannot, do with them. Preferably this thinking occurs in advance of, or in conjunction with, research design. The researcher should attempt to measure variables in a manner that maximizes the potential for achieving the highest level of measurement possible. For example, no amount of data manipulation will turn religious preference into a ratio variable, but consider the manner in which one might approach the assessment of age. Given the option of asking people to respond to a questionnaire item that forces them to classify themselves into an "age category" as opposed to simply asking them their age at their last birthday, what choice makes the most sense? A question designed as in the following example is easy to respond to and avoids the potential embarrassment of having to report one's exact age.

Which category below best describes your age?
1. Under 25
2. 26-35
3. 36-45
4. 46-55
5. Over 55

Unfortunately, this seemingly simple question forces a ratio variable into an ordinal set of categories. The amount of information lost through this procedure is tremendous, as a variable with a potentially large amount of variability is reduced to one with only five categories. The more restricted the range of ages is, the more "information loss" is involved. For example, if those who responded to the age question were between the ages of 26 and 45, then a potential range of 20 categories would be reduced to 2.

As a second example, consider a questionnaire designed to assess support, or lack of it, for the death penalty:

Are you in favor of the death penalty for those convicted of murder?
1. Yes
2. No

Although this might be considered an ordinal variable (or even an interval variable if we consider the possibility of "dummy coding"), one might be able to make a better case for this position if the question were asked in a manner that allowed the respondent to express his or her opinion along some dimension. For example,

Please indicate your level of support for, or opposition to, the death penalty for those convicted of murder.
1. Strongly opposed
2. Moderately opposed
3. Undecided
4. Moderately in favor of
5. Strongly in favor of

In sum, assumptions about the properties of the measurements of variables underlie the use of statistical techniques. When there is likely to be disagreement about whether or not these assumptions have been met, the researcher should have thought through the issues and should be satisfied that he or she can defend the choice of statistical techniques. We will not discuss this issue further here, but we devote a separate section to the question of the use of supposedly "ordinal" scales with techniques that require interval or ratio measurement, such as regression analysis, in Chapter 8.

? What Is Measurement Error?

One of the problems with Stevens's approach to measurement is that concepts prevalent in the social sciences often are difficult to describe as either "objects" or "events." Rather, we frequently examine such complex concepts as "authoritarianism" and "family cohesion." The question that then becomes paramount is how we can measure such complex concepts. The answer to this question involves a larger set of questions concerning how to assess the reliability and validity of any measuring instrument. These questions encompass a very large body of literature that we cannot hope to cover here; however, we accept the position of Carmines and Zeller (1979) that "unreliability is always present to at least a limited extent" (p. 12) and that "the objective of attaining a perfectly valid indicator, one that represents the intended, and only the intended, concept is unachievable" (p. 13). If we accept this position, then, in the context of this discussion, we are concerned with the assumptions that statistical procedures are likely to require about measurement error in the concepts being analyzed. In the case of many statistical techniques, the assumption is that the variables in question are being measured without error. For example, in regression analysis, it is assumed that the independent variable is measured in this way. According to Kerlinger and Pedhazur (1982, p. 34), when there *is* measurement error in the independent variable, as measured by the reliability coefficient, the regression coefficient will be underestimated in relation to this reliability. Kerlinger and Pedhazur make a similar statement regarding measurement errors in covariates in ANCOVA (analysis of covariance): "The consequence of using a fallible covariate in ANCOVA is an under-adjustment for initial differences among groups. This may have far-reaching implications for conclusions about treatment effects" (p. 523).

These examples essentially address how one **operationalizes** or measures variables of interest in a research study. The careful researcher will attempt to do so in a manner that permits statistical operations necessary to answer the questions that generated the research in the first place. Although we cannot propose to measure without error, we can make attempts to measure using instruments and procedures that achieve as little measurement error as possible.

? Explicit Recognition of the "Measurement Model"

To answer the question "How do we know if our measures contain error?," the social scientist typically turns to conventional measures of reliability and validity. For example, a reliability coefficient of .8 is often taken as evidence of having achieved a level of reliability acceptable for serious research. In addition to calculating measures of reliability and validity to provide evidence that we have adequately measured the concepts in question, some statistical models allow the integration of both the theoretical (or structural) model and the "measurement model." This work is primarily attributed to a series of ground-breaking papers by Karl Jöreskog and Dag Sörbom (for a compilation, see Magidson, 1979). Jöreskog and Sörbom developed methods for the analysis of causal relations among variables that explicitly included both the relationships between variables and their measurement. This method generally has become known as structural equation modeling. Although procedures such as multiple regression contain very restrictive assumptions regarding measurement errors, Jöreskog's method, which he terms LISREL (for Linear Structural Relations), allows the researcher to include the relations between the unobserved (or latent) variables of interest and the observable indicators of that variable. Rather than assume that variables are measured without error, and proceed with path analysis using multiple regression methods, the researcher using the LISREL framework can make these assumptions explicit and test their reasonableness. The methods for doing this are far too complex to consider here. We strongly recommend a chapter by Carmines and McIver (1981) for an introduction to the LISREL methodology.

WHAT ARE THE ASSUMPTIONS
ABOUT THE STATISTICAL MODEL?

Whenever one conducts a statistical test, one is also implicitly testing a statistical (i.e., mathematical) model. Much of what we have had to say above has in fact involved explicitly stating the characteristics of the model being tested. For example, whenever we conduct statistical tests that involve correlation and regression, we are applying a "linear model" to the data. If the population parameters are not linearly related, then the linear model is inappropriate and should not be used. Part of the difficulty in selecting a statistical test may in fact result from the difficulty in selecting an appropriate model when the variables involved cannot be assumed to be linearly related.

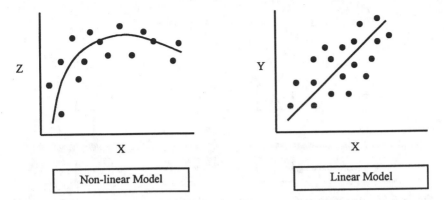

Figure 5.10. Two Scatterplots Representing Different Underlying Models

A simple example might help explain what we are getting at here. Figure 5.10 represents two scatterplots of data with equal variability about their regression lines, and relationships of similar strength between X and Y, and between X and Z. Testing those relationships with the same statistic, however, would not be justified. The bivariate scatterplot on the right represents a linear relationship, whereas the scatterplot on the left represents a relationship that is non-linear. If the researcher did not first examine the bivariate scatterplots and simply ran the Pearson product-moment correlation, the incorrect conclusion would be that X is a strong predictor of Y, but not of Z. Said another way, one of the assumptions that we must be able to make about all our statistical analyses is that we have selected a statistical model that is compatible with our data.

The assumption that two variables are linearly related is one component of a broader or more general assumption known as the assumption of no **specification error**. What this means is that the explanatory model that we have outlined is essentially correct, and that no variables that constitute relevant explanatory factors have been left out, and no variables that are irrelevant have been included. In the broadest sense of the term, specification error refers to any violation of any of the assumptions that underlie a statistical model. In the more narrow usage, specification errors refer to the three kinds of errors we have mentioned so far:

1. Specifying a linear model when the model actually is curvilinear,
2. Excluding relevant variables from the equation, and
3. Including irrelevant variables in the equation.

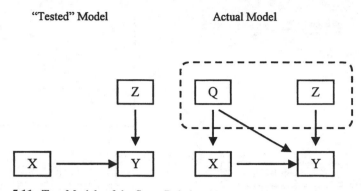

"Tested" Model   **Actual Model**

Figure 5.11. Two Models of the Same Relationship Illustrating Specification Error

Specification errors may seriously bias statistical models. For example, consider once again the bivariate regression equation, containing an error term ε. Recall that ε was said to include all variables, other than X, that influence the dependent variable. This situation is represented in Figure 5.11, which shows two models, one the tested model and one the actual model. The tested model shows two variables, X and Z, that influence Y, the dependent variable. The actual model, however, includes an additional variable, Q, that influences both X and Y.

This variable would be included within ε, but if the variable is correlated with X, then ε and X would also be correlated, thus violating the assumption that error terms remain uncorrelated with independent variables (X).

WHEN SHOULD I BE CONCERNED ABOUT
MEETING THE ASSUMPTIONS OF A TEST?

By now we hope that you are reasonably convinced that the assumptions that underlie statistical analysis are an important piece of the whole data analysis puzzle. If you are convinced, then how should we proceed if the assumptions are not met? If you know what the assumptions of a test are, in many cases, as we have illustrated, it is not a difficult task to examine their reasonableness (unfortunately, in some cases it *is* a fairly difficult task). Our recommendation is that the researcher examine each of the assumptions underlying a statistical test using a general data inspection framework known as exploratory data analysis, or EDA.[1] The philosophy of this approach argues that data analysis should begin with a thorough examination of the structure of one's data, because this provides the researcher with an overview of the variables in question and a starting point for more complex analyses. EDA also makes extensive use of graphic presentations that assist with the visualization of data, under the assumption that both graphic display and numerical summaries are important components of the statistician's toolbox. Three statistical program packages are particularly useful for this purpose: Stata®, Systat®, and SPSS® for Windows®. We have used this approach to illustrate the testing of assumptions underlying data analysis in Chapter 2, and we carry this discussion forward in all of Part III.

NOTE

1. Behrens (1997) provides an excellent introduction to exploratory data analysis. For a more extensive treatment, see Hoaglin et al. (1983).

CHAPTER SIX

How Do I Select the Appropriate Statistical Test?

How do I . . . ? When should or shouldn't I . . . ?

Select the appropriate statistical test?

Conceptualize the relationships between variables in an analysis?

Conceptualize my major research questions for statistical analysis?

Control certain variables?

Use multivariate analysis?

Use correlation and regression analysis?

Use analysis of variance?

Use *t* tests?

Use analysis of covariance?

Use multivariate analysis of variance?

Use correlations?

Use crosstabulations?

What is . . . ? What are . . . ?

A discrete variable?

A continuous variable?

An orthogonal relationship?

Orderable discrete variables?

Covariates?

Level: Intermediate

Focus: Instructional

How Do I Select the Appropriate Statistical Test?

If you already have a basic understanding of statistical analysis, you are probably ready to address the issue of what statistics to use and when to use them. There are a number of published strategies for doing this. These include flow charts, box diagrams, and "expert systems" based computer programs. It is our belief that the selection of an appropriate analytic strategy develops in conjunction with a research design and must be consistent with the type of data collected. When we are asked "What statistic should I use?," our answer usually takes the form of another question: "What question(s) are you trying to answer?" Before presenting models that may be helpful in the selection of an appropriate statistic, we suggest you ask yourself the questions provided in the headings within this chapter.

HOW DO I CONCEPTUALIZE THE RELATIONSHIPS BETWEEN VARIABLES IN AN ANALYSIS?

To answer this question, one must have an idea of the variables of interest and the covariational or causal connections between them. For example, a direct

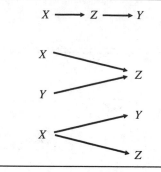

Figure 6.1. Ways to Conceptualize Three-Variable Causal Relationships

causal connection between X and Y, with X independent and Y dependent, is presented in the following causal diagram:

$$X \rightarrow Y$$

One of the easiest ways to clarify your thinking about how variables are related is to construct a diagram like the one above. For example, if your question was, "What is the relationship between number of hours of television viewing and academic performance?," you would probably argue that high levels of TV watching result in or lead to lower academic performance and construct your diagram as follows:

$$\text{TV Viewing} \rightarrow \text{Academic Performance}$$

The type of diagram shown above can be elaborated upon not only to improve your thinking regarding the relationships you may be investigating, but also to assist with the selection of statistical techniques. For example, if your model contained three variables, you might need to decide which of the patterns shown in Figure 6.1 best represented your thinking about the relationships among these variables.

The first example contains an intervening variable, the second suggests that both X and Y contribute to Z, and the third suggests that both Y and Z are caused by X. The point is that before you can conduct a statistical analysis, you must "specify" the patterns of relationships among the variables that interest you.

? What Are Continuous and Discrete Variables?

An important consideration is whether your variables are conceptualized as continuous or discrete. Some statistical analyses will be most appropriate for discrete variables and others for continuous variables. A **discrete variable** is one that classifies the cases or observations into a small number of distinct categories. For example, gender, religious preference, and political preference are all examples of discrete variables. There is no logical order to the categories of these variables, and if the categories were assigned numbers (e.g., 1 = male, 2 = female), these numbers would have no meaning other than to represent the category. Some discrete variables may be ordered. These **orderable discrete variables** most typically are represented by attitude measures on questionnaires. The respondents are required to "strongly agree," "agree," "disagree," or "strongly disagree." Later (see Chapter 8) we discuss the use of orderable discrete variables in statistical analyses that appear to require the use of continuous variables. **Continuous variables**, at least in theory, do not have a minimum-sized unit of measurement. For example, weight is a continuously distributed variable and has no minimum sized unit. No matter how accurately we weigh something, we can always conceive of weighing it just a little better (10.5 lb., or 10.52 lb., or 10.5267193 lb.). Continuous variables do have a range, and each case receives a score within this range. At the very least, continuous variables represent a rank ordering, with larger values meaning more of the property in question than smaller values. In the case of scales constructed for the social sciences, some disagreement may exist regarding whether or not these can be represented as "continuous" distributions. We consider this issue in more depth in Chapter 8.

? When Do I Control Certain Variables?

It is not unusual in research to discover that there are variables whose influence you want to remove or control before examining the effect of an independent on a dependent variable. These variables are sometimes referred to as **control variables** or **covariates** because they "covary" with the independent and dependent variables and therefore may confound the analysis. For example, imagine you are interested in the effect of weight training on gain in muscle mass. If you have two training groups that utilize different training strategies, but one group has 60% females and one has only 25% females, you probably would want to control for sex differences. Thus, sex would become a covariate in this analysis. One way to do this would be to simply analyze the males and females separately. Another would be to use **statistical controls** available with most software packages. In the data analysis selection tables that follow, we include techniques that permit the utilization of covariates in the analysis.

? Are You Comparing Groups or Examining Relationships?

A research question that begins with "What is the difference between . . .?" usually addresses the question of group differences. For example, what differences exist in the perception of competency between those hired through affirmative action programs and those not hired through affirmative action programs (Summers, 1991)? "Group difference" questions typically are those that involve **discrete independent variables** and require the comparison across categories of these variables to answer the research question.

A research question in the form "What is the relationship of X to Y?" usually addresses the question of relationships between variables. For example, Page (1990) investigated, "What is the relationship of shyness and sociability to illicit substance use?" "Relationship" questions typically are those that involve "orderable discrete" or "continuous" independent variables and rely primarily on the calculation of measures of strength of relationship to answer the research question.

? What Other Kinds of Questions Might Research Address?

Research might address a number of other kinds of questions, and we cannot present them all here. Studies concerned with developing or refining existing measurement scales often address complex patterns of relationships among the items composing a scale, all of which are considered dependent variables. These questions typically are addressed with factor analytic techniques or principal components analysis. For example, if we were attempting to develop a scale of marital intimacy (McMain, 1993), we might begin by constructing 100 questionnaire items that assess marital intimacy, then ask the question, "In what way are these items interconnected, and can these interconnections be related to the concept of marital intimacy?"

Some studies attempt to predict to which of several groups a person belongs. For example, Crane (1988) attempted to discover the factors that predict the decision to postpone childbearing. Beginning with two groups of women—those who had made the decision not to have children and those who had made the decision only to postpone childbearing—she conducted a discriminant function analysis that isolated those factors that discriminated the two groups of women. (One of the things that Crane found was that career-related issues formed the major set of variables that discriminated these two groups of women.)

WHEN SHOULD I USE MULTIVARIATE ANALYSIS?

Univariate and bivariate tests refer to the analysis of one or two variables at a time. This generally limits the researcher to bivariate correlation and regression, one-way ANOVA, t tests, a variety of tests based on crosstabulations, and nonparametric tests. Multivariate tests permit the simultaneous analysis of several independent and/or dependent variables. We recognize that in the real world of naturally occurring phenomena, many demographic and situational variables are operative at a given time; thus, the tendency is to utilize techniques that make possible the analysis of many variables simultaneously. In the past, the mathematical operations required to analyze these events were inordinately difficult and time-consuming. The computer revolution has effectively simplified these analyses because of the number-crunching capacity of personal computers and the availability of statistical packages such as SPSS®, SAS®, and SYSTAT®, which continue to become both more powerful and easier to use. This makes it possible to conduct analyses that are extremely complex, both mathematically and conceptually, utilizing dozens of variables simultaneously. Unfortunately, it also makes it possible to conduct inappropriate analyses, with insufficient samples that lead to completely incorrect conclusions.

At the most elemental level, **multivariate** refers to looking at the relationships among more than two variables (one sample t tests and simple correlations are examples of univariate and bivariate techniques, respectively). Most statisticians, however, reserve the term *multivariate analysis* for situations in which there are multiple independent and/or dependent variables and the measured variables are likely to be dependent upon each other (i.e., to correlate). Of course, the impact of each variable can be evaluated independently, but such an evaluation is likely to be confounded by the influence of other variables, hence the desirability and power of multivariate techniques. Thus, multivariate analysis allows for the examination of two variables while simultaneously controlling for the influence of other variables on each of them.

When two variables (X_1 and X_2) correlate with a third variable (Y) but do not correlate with each other, they are said to be **orthogonal**. This might be the case in an experimental study using two independent variables, for instance, length of training (X_1) and amount of humor (X_2) in a driving instruction seminar, and one dependent variable, let's say, changes in driving behavior (Y_1). The design, if truly experimental, is such that there is no overlapping variance (i.e., no correlation) between X_1 and X_2, suggesting that they make independent contributions to the dependent variable. These contributions can be summed to assess the overall effect of the two variables on the dependent variable.

The situation described above is somewhat atypical. In the more typical situation, the independent, or X, variables will correlate. That is especially the case with naturally occurring phenomena that are not experimentally controlled and in cross-sectional survey research. It is no accident that multivariate techniques emanated from nonexperimental research concerns. Thus, the intersection among X_1, X_2, and Y represents the overlapping variance that contributes to the overall relationship but cannot easily be attributed to either X variable. (The technical way to say this is that the variability cannot be uniquely partitioned.) An example might be the impact of reading ability (X_1) and time spent studying (X_2) on material learned (Y). Here, it is likely that there is a nontrivial correlation between reading ability and study time.

Whenever there is overlapping variance, the researcher has some choices regarding how to analyze the data. One strategy is to disregard the overlapping variance and evaluate the impact of the independent variables simultaneously. This leads to a solution in which the researcher gains an understanding of the overall contribution of the variables to the dependent variable, and the unique contribution of each variable, having removed or controlled the overlapping variance. Another strategy, called hierarchical analysis, consists of entering the variables into the analytical framework in some order that gives priority to one variable (the first one entered) by assigning that variable both its unique variance and any overlapping variance. These particular strategies are described more fully in Chapter 10, which covers multiple regression analysis.

Multivariate analyses are particularly useful for predicting scores or group membership. They do so by creating linear combinations of variables with empirically determined weights, as shown in the following equation:

$$\text{Value of dependent variable} = a_1X_1 + a_2X_2 + a_3X_3 + a_4X_4 + \ldots + a_nX_n$$

where:

a_i = the weight assigned to variable i
X_i = an independent variable.

The equation above signifies a data set consisting of values on several measures for several subjects. Those variables designated as independent variables are "weighted" as a function of their contribution to the prediction of the dependent variable. (The Xs in a multivariate solution could also be cross-products, which indicate interactions between predictor variables [e.g., X_1X_2] or single variables raised to a power [e.g., X^n].)

The domain of multivariate techniques is expanding. Included in this domain are the most commonly referenced statistical methods: multivariate analysis of variance and covariance, multiple regression, multiple discriminant function analysis, factor analysis, cluster analysis, multidimensional scaling, structural equation modeling, logistic regression and logit and probit analysis, and many others, including a wide variety of time-series analyses. There is a certain logic to selecting the appropriate technique, and our goal is to describe this logic in this chapter; however, you will need to consult more advanced statistics books for detailed information about these complex methods of analysis. As in all instances of research design, the research problem, whether theoretical or applied, must first be examined at a conceptual level before determining the variables and ways to analyze their relationships.

Multivariate techniques frequently are used inappropriately, even by experienced researchers. The reasons for this include their ease of calculation, an incomplete understanding of the assumptions underlying their use (see Chapter 5 for a discussion of the assumptions of statistical testing), and a failure to examine the structure of the data prior to conducting the analyses (see Chapters 2 and 7). We find, however, that many of the problems with the use of multivariate analysis lie in the fact that the sample was simply too small. Even though multivariate methods generally are desirable, they work better with large samples. With small samples, the estimates produced by these methods are likely to be unreliable and/or biased. Based on the information above, we offer the following suggestions regarding the use of multivariate techniques.

1. Plan the statistical approach in advance. If multivariate analysis is called for, determine the number of cases necessary and calculate the power that this sample will provide.

2. Be aware of the assumptions of the multivariate method and examine the data to acquire an understanding of how well these assumptions have been met.

3. Build multivariate models from the more simple to the more complex. Start with an examination of the bivariate building blocks of the multivariate model.

4. Seek to acquire a complete understanding of both the overall effects of a set of independent variables and the role of each independent variable within that set.

5. Remember that these analyses are complex and are rarely a "one-shot" process. Think of the analyses as an iterative, trial-and-error exploration of your data.

HOW DO I SELECT THE APPROPRIATE STATISTICAL TEST?

In this section, we suggest a framework for the selection of an appropriate statistical technique. The framework does not include all possible designs but

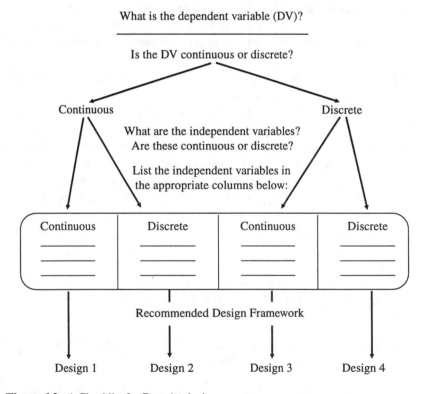

Figure 6.2. A Checklist for Data Analysis

can be viewed as a starting point for thinking about the selection of a statistical technique. We ask you first to fill in a short checklist (Figure 6.2) regarding the characteristics of your data analysis problem and then point you to a design framework that addresses that problem.

? Major Research Questions Suggested by the Four Design Frameworks

Design 1: What is the degree or strength of relationship between the dependent variable and independent variable(s)?

Design 2: Are there significant group differences between the groups formed by the independent variable(s) and scores on the dependent variable?

Design 3: Are the scores on the independent variables significantly related to the categories formed by the dependent variable? Can the scores or values of the independent variables discriminate groups, or predict group membership?

Design 4: Are differences in the frequency of occurrence of the independent variable related to differences in the frequency of occurrence of the independent variables?

? Statistical Analyses Suggested by Each Design Framework

We have divided the question of selecting an appropriate statistical technique into four "designs" or "design frameworks." These design frameworks are based on the characteristics of the dependent and independent variables and lead to, or are based on, a specific type of research question. Below we reintroduce each of these, along with providing a short introduction to help identify the appropriateness of that design for a particular research question.

Design 1

Use this design if your independent variable or variables represent continuously distributed data and your dependent variable is also continuously distributed.

Research Question: What is the **degree or strength of relationship** between the dependent variable and independent variable(s)?

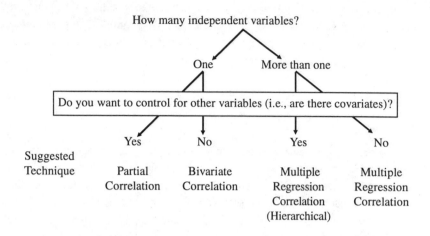

Note: If there are multiple dependent variables, either (a) analyze each independently, (b) use canonical correlation, or (c) utilize path analysis or structural equation models.

Design 2

Use this design if your independent variable or variables represent discrete categories that form groups that you want to compare and your dependent variable is continuously distributed.

Research Question: Are there significant group differences between the groups formed by the independent variable(s) and scores on the dependent variable?

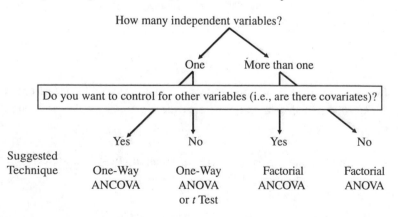

Note: ANOVA = analysis of variance, ANCOVA = analysis of covariance, MANOVA = multivariate analysis of variance, and MANCOVA = multivariate analysis of covariance. If there are multiple dependent variables, without covariates, utilize MANOVA; if there are covariates, utilize MANCOVA.

Design 3

Use this design if your independent variable(s) are continuously distributed and your dependent variable is discrete.

Research Question: Are differences in the frequency of occurrence of the independent variable related to differences in the frequency of occurrence of the dependent variables?

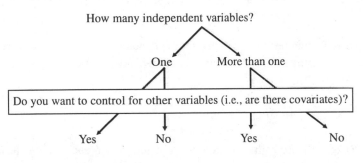

Suggested techniques: logistic regression or discriminant function analysis (hierarchical if covariates are present). If there are more than two categories of the dependent variable, use multiple discriminant functions or polytomous logistic regression analysis. In the case of one independent variable, the results will be similar to the results of one-way ANOVA. If there are multiple dependent variables, use factorial discriminant function analysis or analyze each dependent variable independently.

Design 4

Use this design if both your independent variables and dependent variables are discrete.

Research Question: Are there significant group differences between the groups formed by the cross-classification of the independent variable(s) and the dependent variable?

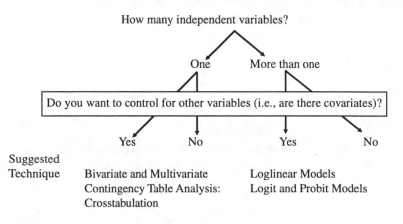

How many independent variables?

One   More than one

Do you want to control for other variables (i.e., are there covariates)?

Yes   No   Yes   No

Suggested
Technique   Bivariate and Multivariate   Loglinear Models
                 Contingency Table Analysis: Logit and Probit Models
                 Crosstabulation

Note: Techniques to deal with multiple dependent variables are not available. Analyze multiple dependent variables independently or use as controls with other dependent variables.

? Describing the Techniques

Boxes 6.1 through 6.5 provide descriptions of each of the techniques mentioned in the four designs above and suggest sources for further reading.

Box 6.1

PEARSON PRODUCT-MOMENT CORRELATION
AND BIVARIATE REGRESSION ANALYSIS

Independent variables	One, continuously distributed
Dependent variables	One, continuously distributed
Null hypothesis	$H_0: \rho = \beta = 0$
Statistical tests	F for bivariate correlation/regression or Fisher's r to z transformation
Measures of association	Pearson's r or r^2, ranges from -1 to $+1$; zero indicates no relationship
Measures of effect size	Use r: small = .10, medium = .25, large = .45
Controls/covariates: how handled	Use partial correlation to control effect of extraneous variables or correlate variables of interest in subgroups of control variable
Comments	These techniques assume a linear relationship between variables and bivariate normality
References	Beginning: Any introductory statistics text with social science applications Intermediate/advanced: See references in Box 6.2

Box 6.2

Multiple Correlation and Regression Analysis

Independent variables	Two or more, continuously distributed (but may use dummy coded categorical independent variable)
Dependent variables	One, continuously distributed
Null hypothesis	H_0: $\rho = 0$, $\beta_i = 0$
Statistical tests	F for multiple correlation, F or t for regression coefficients
Measures of association	Use multiple R or R^2 and standardized regression coefficients interpreted in comparison with other independent variables
Measures of effect size	Use multiple R or R^2 or convert to effect size (ES) using the formula ES $= R^2/(1 - R^2)$: Small $= .02$, medium $= .15$, large $= .35$.
Controls/covariates: how handled	Use hierarchical multiple regression to remove effect of extraneous variables
Comments	These techniques assume a linear relationship between independent and dependent variables and multivariate normality; dependent variables may not be dummy coded
References	Beginning: Bohrnstedt and Knoke (1994), Howell (1997), Kachigan (1991) Intermediate: Hamilton (1992), Kerlinger (1986), Tabachnick and Fidell (1996) Advanced: Cohen and Cohen (1983), Cook and Weisberg (1994), Kerlinger and Pedhazur (1973), Sage series in regression and correlation for individual topics

Box 6.3

ONE-WAY ANALYSIS OF VARIANCE AND T TESTS

Independent variables	One, discretely distributed (dichotomous for t tests)
Dependent variables	One, continuously distributed
Null hypothesis	$H_0: \mu_1 = \mu_2$ (t tests) $H_0: \mu_1 = \mu_2 = \mu_3 = \ldots = \mu_n$ (F tests)
Statistical tests	F or t for dichotomies, F for one-way ANOVA
Measures of association	Size of mean difference, t converted to r, eta and eta^2 for one-way ANOVA
Measures of effect size	Standardized difference between means ($[M_1 - M_2]/SD_w$): small = .20, medium = .50, large = .80
Controls/covariates: how handled	None for t tests, one-way ANCOVA for F tests; if independent variable is dichotomous, use one-way ANCOVA in lieu of t
Comments	For dichotomous independent variable, $t^2 = F$
References	Most introductory texts, particularly those from a psychological perspective; considered a preliminary topic in more advanced texts

Box 6.4

FACTORIAL ANALYSIS OF VARIANCE

Independent variables	More than one, discretely distributed
Dependent variables	One, continuously distributed
Null hypothesis	H_0: $\mu_1 = \mu_2 = \mu_3 = \ldots = \mu_n$ (F tests)
Statistical tests	F, varies for a priori and post hoc comparisons
Measures of association	Size of mean differences, eta and eta^2
Measures of effect size	Eta^2; small = .10, medium = .25, large = .45
Controls/covariates: how handled	ANCOVA
Comments	Planned and post hoc comparisons discussed in Keppel (1991) and Kirk (1995)
References	Beginning: Kachigan (1991) Intermediate: Keppel (1991), Tabachnik and Fidell (1996) Advanced: Kirk (1995), Myers (1979), Winer (1971)

Box 6.5

CROSSTABULATION: BIVARIATE AND
MULTIVARIATE CONTINGENCY TABLE ANALYSIS

Independent variables	One or more, discretely distributed
Dependent variables	One, discretely distributed
Null hypothesis	H_0: $\chi^2 = 0$, zero association defined by randomness in distribution of cases in cells
Statistical tests	Chi-square test
Measures of association	Many: odds ratio and log odds, phi, Cramer's V, contingency coefficient, lambda, gamma, tau, Somer's D, and others
Measures of effect size	Many, no one standard, see above
Controls/covariates: how handled	Multivariate contingency table analysis
Comments	These techniques work best with a small number of discrete variables; more advanced techniques are used with greater numbers of variables or a mixture of continuous and discrete independent variables
References	Beginning: Babbie (1997), Nachmias and Nachmias (1995) Intermediate: Bohrnstedt and Knoke (1994), Jacobson (1976)

PART three

Issues Related to Variables and Their Distributions

Up to this point, we have discussed the preliminary steps that a researcher must take in preparing data for statistical analysis. These steps include organizing the data for analysis and selecting the appropriate statistical tests. One can learn a great deal at this stage by taking the time to visually inspect the collected data prior to submitting them to inferential statistics. Visual inspection is also a recommended way of circumventing problems that might arise in trying to interpret subsequent statistical analyses.

The chapters in Part III are responses to frequently asked questions about data. Chapter 7 discusses the implications of missing data, extreme data (outliers), and data that are not normally distributed. These issues become critical in a number of contexts related to statistical analysis. Some scores may be so distant from the middle of the distribution that they appear to be aberrations. These outliers tend to skew data distributions, and thus may produce distributions that violate the assumption of normality required for much statistical testing. Missing data raise questions about the adequacy of the research process or measuring instrument and may lead to violation of the assumption of random selection, also required for statistical testing. Finally, the distribution of scores itself may look quite odd, not at all like the normal distribution on which statistical inference theory is based. Chapter 7 highlights these controversies, offers advice on assessing the extent of potential problems, and provides recommendations for solutions.

Chapter 8 addresses three additional frequently asked questions about the treatment of variables in statistical analysis. The first question, a very common source of disagreement among data analysts, concerns questions surrounding level of measurement. When can a variable be considered intervally scaled, and what statistics are appropriate when the assumption of equal intervals is not warranted? Second, Chapter 8 addresses the question of "dummy variables" and their use in statistical analysis, and the related issue of the appropriateness of creating dummy variables from continuous distributions.

To benefit from the content of these two chapters, it is expected that you already have a basic understanding of data analysis terminology and procedures. The material in Chapter 7 is important to anyone at the early stages of assessing the form and adequacy of the data from a research study prior to conducting inferential tests. The first part of Chapter 8 is also relevant to that stage of statistical inquiry. The remainder of Chapter 8 addresses issues that demand a somewhat more sophisticated understanding of statistical methods.

CHAPTER seven

How Do I Deal With Non-Normality, Missing Values, and Outliers?

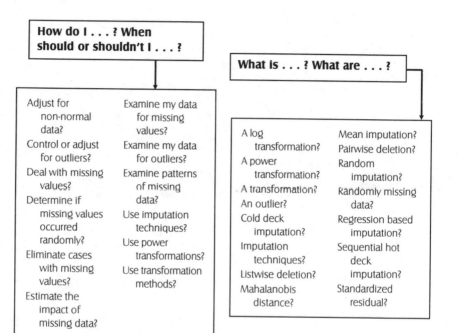

How do I . . . ? When should or shouldn't I . . . ?

Adjust for non-normal data?

Control or adjust for outliers?

Deal with missing values?

Determine if missing values occurred randomly?

Eliminate cases with missing values?

Estimate the impact of missing data?

Examine my data for missing values?

Examine my data for outliers?

Examine patterns of missing data?

Use imputation techniques?

Use power transformations?

Use transformation methods?

What is . . . ? What are . . . ?

A log transformation?

A power transformation?

A transformation?

An outlier?

Cold deck imputation?

Imputation techniques?

Listwise deletion?

Mahalanobis distance?

Mean imputation?

Pairwise deletion?

Random imputation?

Randomly missing data?

Regression based imputation?

Sequential hot deck imputation?

Standardized residual?

Level: Intermediate

Focus: Instructional

How Do I Deal With Non-Normality, Missing Values, and Outliers?

Chapter 7 is a response to three frequently asked questions about data. The first section discusses the implications of missing data. What do you do when some items of information are unobtainable or are mysteriously missing from your data pool? There is a certain amount of controversy about the treatment of missing data, and the answers range from the simple to the complex. The second section confronts the issue of outliers, those data points or responses from research participants that seem to lie outside the range of the bulk of the data. Here too, there are many possible solutions. Our recommendations rest on making an adequate assessment of the nature and seriousness of the problem. Finally, we address the question of what to do with data that are not normally distributed. This issue was broached initially in Chapter 5 in the context of the assumptions that provide the foundation for statistical testing. In practice, however, data do not always behave as they do in theory. When violations to the assumption of normality warrant special procedures, the researcher needs to know what to do. In this chapter, we offer our suggestions.

HOW DO I DEAL WITH MISSING VALUES?

As we gather data for each case in a research endeavor, it is almost a certainty that some items of information will be unobtainable. These incomplete or lost responses are referred to as "missing values" and result in an incomplete set of data for some cases. There are innumerable reasons for missing values in research. Following is a list of only a few of the possible explanations:

The respondent refused to answer one or more questions on a survey.

The respondent is not home when the interviewer calls.

During a phone interview, the respondent hangs up.

An interviewer inadvertently skips one or more questions.

An experimental animal dies halfway through the study.

A recording instrument fails.

A survey is collated improperly, and a page is left out of some surveys.

Data collected by other persons or organizations—the U.S. Census, for example— may have certain pieces of information not available for every case.

Missing data are an unavoidable reality in research. The implications of missing data include the possibility of making inferences on the basis of sample data that are inadvertently biased in unknown directions, as well as being forced to rely on reduced sample sizes for statistical analysis. What does the researcher do? Ignore missing data? Fill in arbitrary values to complete the distribution of scores?

The nature of statistical inference, generalizing from sample observations to conclusions about populations, presupposes that data are missing; that is, the sample differs from the population precisely because it does not contain all the observations within the population. The logic of statistical inference, however, presumes that the sample is **randomly** drawn from the population. Similarly, whether or not the missing data within a sample are random is an important consideration. Randomly missing data means that there is no systematic difference between the available data and the missing data; they are both random subsets of the data composing the entire sample. There is no acknowledged way of making this determination, but there are guidelines available to the researcher. In practice, we become suspicious if a large amount of data are missing from a certain variable, because it cannot be assumed that the missing data are representative of the remaining data. The missing values may imply, for example, that certain types of subjects had difficulty with an item. Excising the scores of these subjects would be prejudicial. For example, some people may be very willing to answer questions about their sexual activity, whereas others may be equally unwilling to do so. These are most likely very different sorts of people who probably are very different on a number of other variables related to their attitudes and values about sexuality and morality.

As a first step, the researcher seeks to determine the extent of the missing data in a data file. The procedures suggested for the examination of data in the second chapter will make the problem very clear. Note that in some cases, data *should* be missing, as when a question is the result of a filter from a previous question. For example, it makes no sense to ask people the age at which they were married when a previous question has established that they have never been married. All single, never married persons should have data missing on such a question.

The researcher first seeks to determine the reasons behind the missing data. That will help determine if the missing data are the result of random oversights or systematic bias. Sometimes, data are missing because of data collection or data entry problems. More frequently, data are not available on particular variables from certain research participants, or the respondents themselves neglected to answer or refused to answer especially taxing or intrusive questions. In the latter instance, you may be able to avoid having substantial missing data by designing questions that are more benign or by collecting the data more sensitively. For example, rather than ask people directly what their annual income is, survey researchers often hand respondents a card with income categories represented by letters and ask the respondents to give the letter of the category that comes closest to their annual income. Thus, a gauge of annual income can be obtained without the respondents ever having to provide a direct dollar figure.

If a review of the research procedures and an overview of the missing data are not sufficient to identify distinct patterns, more formal steps may need to be taken to make this determination. If you cannot be sure that the missing entries are random, then any statistical results based on these data will be biased if the distributions of included variables are affected by the missing data.

One method to determine if the process resulting in missing data is a random one is to form two groups, one consisting of cases that contain missing values on a particular variable and the other group consisting of cases without the missing values (Hair et al., 1995). These groups are then compared for patterns of significant difference. If these patterns are found, they suggest a nonrandom process of missing data acquisition. For example, let's say we are conducting a survey on the number of sexual partners acknowledged by men and women. Not surprisingly, some participants leave this question unanswered. We now compare the percentage of men and women who answered the question on sexual partners with those who left it blank. If these percentages are almost equal, it appears that the data are randomly missing; if these percentages are significantly different, the data are not randomly missing, at least with respect to the variable of gender. With continuous variables rather than categorical variables, the comparison would be made using a t test rather than percentages. Note, however, that drawing conclusions about randomly missing data on the basis of nonsignificant differences between groups is akin to confirming the null hypothesis, a practice we took issue with in Chapter 3.

A second method uses dichotomized correlations to evaluate the correlation of missing data between variables (Hair et al., 1995). With this approach, all valid values of a variable are coded with the value of one, and all missing data are coded with a value of zero. Using these assigned values, every pair of variables is now correlated. The correlations show the level of association between the missing data for each pair. Randomness within a pair of variables is indicated by low correlations. A statistical significance test of the correlations offers a conservative estimate of randomness. Note that lack of randomness in this context means that the observed values for a variable may still represent a random sample of values for each value of the paired variable, but not necessarily a random sample of all values on that variable. With regard to the previous example, missing data on prior sexual partners might occur randomly among both men and women, but much more frequently for women than men. Consequently, any remedial procedure to accommodate the missing data on sexual partners must consider the gender of the respondent.

After the researcher obtains a clearer understanding of the scope of the problem, a number of alternative procedures can be instituted to deal with it. If data are believed to be missing randomly, there are a number of techniques for intervening. One is to simply delete **subjects** with missing data. This is the default option in many computerized statistical programs. It is a straightforward method but has the significant disadvantage of reducing power through subject loss. Only in cases where a few subjects account for a substantial amount of missing data, or where there are a large number of subjects available and very little missing data, is this strategy recommended. An exception, noted by Hair et al. (1995), is to delete cases where the missing values occur in the dependent variable of a statistical analysis. In general, however, whenever missing values are distributed throughout cases and variables in a multivariate study, deletion of entire cases leads to considerable loss of data. When the data are organized in an experimental design, losing a single case may result in unequal cell sizes and lead to more complicated data analyses. We have found that in research, no matter how carefully designed the data collection instruments or how carefully the potential participants are screened, there typically will be a few cases in which the instruments are largely incomplete. When this is the case, we typically assume that the respondent was not interested, was unmotivated, or worse yet, made a conscious attempt to sabotage the research. We recommend eliminating all data for that case.

A related approach is to eliminate **variables** that are associated with considerable data loss from the study. This may be the most efficient approach to dealing with a nonrandom pattern of missing data. It also may be unavoidable because items that are left unanswered by many subjects are likely to be untrustworthy. On the other hand, no investigator wants to lose key variables from a study. One rule of thumb suggested by Hertel (1976) is that a variable should be excluded from analysis if 15% or more of subjects are missing data on it.

A second major approach to handling missing values is to use "imputation" techniques. **Imputation** refers to estimating missing values and then using the estimates in subsequent statistical analyses. Imputation is not used as frequently for overcoming the limitations of missing data as the previously cited techniques because it requires more expertise, the computations can be difficult, and statistical program packages do not make imputation easily accessible to the user. Recently, SPSS Inc. developed "SPSS Missing Value Analysis®," a program that assesses the magnitude of each pattern of missing data within a table. According to the published literature available from the SPSS website (www.spss.com),

The SPSS Missing Values Analysis module provides two methods for maximum likelihood estimation and imputation. First, the EM (Expectation-Maximization) algorithm is an iterative algorithm that can provide estimates of statistical quantities such as correlations, or imputed values for missing values, in the presence of a general pattern of missingness. Second, regression imputation relies on the fact that the EM approach is mathematically very similar to using regression to fill in the missing values using predicted values from a regression of a given variable on other variables in the analysis. Imputing regression predictions in this fashion can underrepresent the variance of the variable in question, so one might "jitter" the predicted values by adding a random component to the values. Either or both of these methods can be tried on your data using SPSS Missing Values Analysis. You can inspect the results, and in general you expect them to perform similarly. In general, either of these methods are superior to approaches such as listwise deletion, pairwise deletion, or mean substitution.

As indicated above, many forms of imputation are available, including complex model-based procedures that can be employed with nonrandom missing data (Little & Rubin, 1987). We will address only common ones here. Each approach can be implemented either by using data from observations with no missing data to estimate missing data on the remaining cases (the "complete case approach") or by using data from all available valid observations to make these estimates (the "all-available approach") (Hair et al., 1995).

The most straightforward and commonly used method is **mean imputation** (also called mean substitution), which consists of entering the mean value of a variable for any subject with missing data on that variable. Mean estimation is a conservative procedure, in that the mean of the distribution of that variable does not change. The procedure will, however, artificially reduce the variance of the distribution and thus may reduce the correlation of the variable with other variables. Although the approach results in a reduction of power, it is certainly easy to administer and is generally preferable to making a reasoned guess about the value of the missing variable.

A second technique is known as **random imputation** or **sequential hot deck imputation** (Little & Rubin, 1987). The idea is to replace the missing value with a value randomly chosen from the available cases. The term "sequential hot deck imputation" comes from the process of arranging a data file randomly and utilizing the file adjacent to the file with the missing value to provide that score. This approach does not systematically affect the variance of the distribution in the way that mean imputation does, but it does introduce more random variability. The procedure also can be contrasted with **cold deck imputation**, in which the transferred value is a constant value that comes from previous research or external sources. The cold deck procedure has the same disadvantages as the mean imputation procedure.

A more sophisticated method for estimating missing values is to rely on **regression values**. The idea is to construct a regression equation with the other variables as independent variables and the variable with missing data as the dependent variable. The equation is derived from subjects without missing data and is then used to predict the missing values for the remaining subjects. Typically, the predicted values from the first round of regression are assigned for missing values, and then all the cases are used in a second regression. The predicted values for the variable with missing data from this round are the basis for a third regression. The process keeps going until the predicted values from one round to the next are similar. The predictions from the last round are then chosen to replace the missing values.

The advantage of the regression approach is that it offers a more accurate estimate of missing values. The disadvantages are that it is computationally complex and that scores taken from regressions fit together better than they should because the estimates have been based on the other variables and are likely to be more consistent with them than actual scores would be (Tabachnick & Fidell, 1996). Thus, the method reinforces the relationships present in the sample data, which then become less generalizable.

It is particularly aggravating when the data that are missing consist of outcomes for participants who failed to complete a study or procedure. Ignoring these participants and performing the analysis on those who completed the study not only reduces sample size but also runs the risk of bias, because attrition may be directly related to treatment condition. Most imputation methods also have limitations here. Simple mean imputation, of course, assumes that attrition is random over the entire study, which usually is not the case. More sophisticated methods of imputation, such as those proposed by Pigott (1994), are better, but they still assume that the mechanisms that account for the missing data can be ignored. They may not be completely random, but at least they are not related to the actual values of the missing data, a situation that is more likely to be defensible. Moreover, such methods of imputation assume that there are variables available that are good predictors of the treatment outcomes.

Shadish, Hu, Glaser, Kownacki, and Wong (1998) present a method for dealing with attrition when these conditions are not met. Their method has the advantage of being available as a PC-based computer program called Attrition Analyzer (available at ftp://irvin.psyc.memphis.edu/pub/outgoing/att.exe). It is limited, however, to situations in which participants are randomly assigned to groups, where treatment outcome is a dichotomous variable, and where the outcome status of some participants is unavailable. Examples of relevant studies they cite include smoking cessation research (quit or nonquit), marital research (divorce or nondivorce), and medical research (mortality or not).

TABLE 7.1   Determining Odds Ratios in 2 × 2 Tables

| Condition | Outcome | |
	Success	Failure
Treatment	A	B
Comparison	C	D

Shadish et al.'s (1998) approach makes use of the odds ratio (see Chapter 4), which is the appropriate measure of effect size for fourfold tables such as Table 7.1. The odds ratio in this example is OR = AD/BC. An odds ratio of 1.00 means that participants in the two groups did equally well, whereas an odds ratio that exceeds 1.00 indicates the superiority of the treatment group and an odds ratio of less than 1.00 indicates the superiority of the comparison group. The odds ratio is first computed for all participants whose outcome data are available. The method next computes the range of possible odds ratios that could be observed in the study given attrition. Many different conclusions could be drawn, of course, depending on the successes and failures of the missing participants, but not all these patterns are equally likely. Thus, the technique determines the probability of observing odds ratios at or beyond any particular value, based on how many participants are missing from each condition. One could choose a threshold value of 1.00, for instance, or perhaps the smallest clinically significant odds ratio worth noting, or maybe the smallest possible statistically significant value. Readers whose data conform to dichotomous outcomes with two treatment groups are referred to the Shadish et al. (1998) article and/or website for more information about this ingenious technique for adjusting data to the compromising impact of attrition.

? Missing Values: A Detailed Example

Below, we present a detailed example that describes an attempt to deal with missing values in a small data set. We begin with a 49-case data matrix containing five items from a self-esteem scale. Each item has a 4-point scale as its measure. We first present the mean and standard deviation for each item in Table 7.2.

We next utilize a table of random numbers to randomly delete 29 of the data points. Because there are five variables and 49 cases, this amounts to deleting

TABLE 7.2 Complete Data Set: No Missing Values

Variable	Mean	Standard Deviation	Minimum	Maximum	N
Item 1	2.49	.96	1.00	4.00	49
Item 2	2.18	1.69	1.00	4.00	49
Item 3	1.31	.47	1.00	4.00	49
Item 4	1.71	.79	1.00	4.00	49
Item 5	2.04	.98	1.00	4.00	49

approximately 12% of the observations. The new table of descriptive statistics for the five variables is shown as Table 7.3. Note that the smallest valid N is 41, indicating that eight values are missing for Item 4.

Note that the changes in means and standard deviations from the full data set appear minor. The largest is an increase in the mean value from 2.18 to 2.24 for Item 2. This appearance may be misleading, however. This is because missing data often exert their influence only after the researcher attempts an operation that requires the use of two or more of the variables at once. We give two examples of this below.

? Example 1: Bivariate Correlation Analysis

First, there are two different ways that we could calculate the correlations among these five variables. We could simply correlate each pair, eliminating the cases that have missing values for that pair only (**pairwise deletion**), or we could use only the cases that have complete data for all variables we wish to correlate

TABLE 7.3 Data Set With Randomly Selected Missing Data

Variable	Mean	Standard Deviation	Minimum	Maximum	N
Item 1	2.50	.99	1.00	4.00	42
Item 2	2.24	1.10	1.00	4.00	42
Item 3	1.33	.47	1.00	2.00	43
Item 4	1.71	.81	1.00	4.00	41
Item 5	2.02	1.00	1.00	4.00	44

TABLE 7.4 Three Correlation Matrices Based on Different Missing Value Treatments

Correlation Matrix (No Missing Data; N = 49)

	Item 1	Item 2	Item 3	Item 4
Item 1	1.000			
Item 2	.3691	1.000		
Item 3	.3566	.0510	1.000	
Item 4	.2157	.2790	.4123	1.000
Item 5	.3111	.5385	.1550	.2579

Correlation Matrix (Listwise Deletion; N = 21)

	Item 1	Item 2	Item 3	Item 4
Item 1	1.000			
Item 2	.2571	1.000		
Item 3	.2235	−.0451	1.000	
Item 4	.0275	.1519	.5597	1.000
Item 5	.3495	.3093	.2365	.3338

Correlation Matrix (Pairwise Deletion; N in brackets)

	Item 1	Item 2	Item 3	Item 4	
Item 1	1.000				
	[42]				
Item 2	.4417	1.000			
	[35]	[42]			
Item 3	.3188	.0015	1.000		
	[36]	[36]	[43]		
Item 4	.1983	.1773	.5916	1.000	
	[34]	[37]	[35]	[41]	
Item 5	.4500	.5798	.1832	.2431	1.000
	[37]	[38]	[38]	[38]	[44]

(**listwise deletion**). In Table 7.4, we present three correlation matrices. The first contains the "real" correlations, those calculated before we randomly created missing values. The second contains the correlations based on pairwise deletion, and the third shows the correlations based on listwise deletion.

We begin with an examination of the relationship between Item 1 and Item 2 in each matrix. The "true" value is .3691. The listwise value, .2571 ($N = 21$),

TABLE 7.5 Scale Scores Based on Different Missing Value Treatments				
Scale	*Mean*	*SD*	*N*	*Alpha*
Complete matrix	9.69	2.94	49	.68
Missing data matrix	8.11	2.63	21	.58

is 30% lower, and the pairwise value, based on 35 cases, is .4417, or 20% greater. Even though the differences probably would not be so dramatic with a larger data set and with normally distributed continuous data, this demonstration is very realistic in that it represents the types of data that social scientists are likely to work with and illustrates the point that the manner in which one chooses to deal with missing data can result in very different interpretations concerning the relationships in the data.

We can make a number of additional important points about the above comparisons. First, with listwise deletion, we lose more than 50% of the data, creating an unacceptable loss of power and accuracy. Listwise deletion would not be a satisfactory solution in this case. Second, we cannot predict whether or not the correlations will be inflated or deflated from their original values when we use listwise or pairwise deletion. Some drop considerably, and others are increased. Third, one might come to the conclusion that pairwise deletion is generally more "accurate" because it is based on more cases and, in this example, more closely approximates the original data. This may not always be the case if nonrandom patterns of missing data infiltrate the data set. Remember, however, that with pairwise deletion of missing values, conclusions based solely on bivariate data will be based on different numbers of cases. When patterns of missing values are not random, this may mean that different correlations actually represent different populations.

? Example 2: Calculating Scale Scores

Another operation that a researcher might want to perform with a data set such as this is to sum the values to form a scale. Because the five items actually compose a self-esteem scale, it would be reasonable to add the items to form a five-item scale. In Table 7.5, we present the scale mean, *N*, and standard deviation, along with the alpha reliability for the complete data matrix and for the matrix containing missing data.

TABLE 7.6 Correlation Matrix Based on Dummy Coded Missing and Nonmissing Data

	Item 1	Item 2	Item 3	Item 4	Item 5
Item 1	1.000				
	(49)				
Item 2	**.3654**	1.000			
	(49)	(49)			
Item 3	.2569	.0053	1.000		
	(49)	(49)	(49)		
Item 4	.1564	.1614	.1575	1.000	
	(49)	(49)	(49)	(49)	
Item 5	.1043	.1299	.1644	.2284	1.000
	(49)	(49)	(49)	(49)	(49)

Note that in this case, we would conclude that the average self-esteem score is more than 1.5 points lower than that shown by the complete data set and that the reliability is also lower. Note also that it is highly undesirable to present a scale score based on "pairwise" deletion because we can add only cases that have complete data, and there are only 21 such cases in this data set. This suggests that we need another method to avoid the loss of more than half of our data. We discussed a number of "imputation" techniques above, along with their advantages and limitations. Prior to utilizing one of these methods, it is important to examine the *pattern* of missing values. A number of methods for doing this were also discussed above. We consider one of these techniques below and then utilize the "mean substitution" method of imputation to re-examine the correlation matrix and scale scores.

? When Should I Use Imputation Techniques?

Recall that there are a number of methods of "imputation" that allow a researcher to "substitute" a value to replace a value that is missing. These techniques are most legitimate when we are confident that the pattern of missing data is "random." In the data set we are using, we are confident that the data loss is random because we used a table of random numbers to determine which values

TABLE 7.7 Correlation Matrix: Mean Substitution ($N = 49$)

	Item 1	Item 2	Item 3	Item 4
Item 1	1.000			
Item 2	.3651	1.000		
Item 3	.3481	.0352	1.000	
Item 4	.2015	.2910	.4226	1.000
Item 5	.3364	.5099	.1644	.2611

to assign as missing. We have coded all the missing values for Items 1 to 5 as "1" and coded all valid values as "0" and correlated the variables. The results are shown in Table 7.6.

The pattern of correlations shown in Table 7.6 indicates that only the correlation of Item 1 with Item 2 (.3654, boldfaced) is statistically significant, suggesting that the pattern may not be random as far as these two variables are concerned. We might hypothesize that when a person left one of these questions blank, he or she was also likely to leave the other blank, and then search for similarities in the questions that suggest this pattern. All the other correlations are not statistically significant; however, the reader might realize that with a slightly larger sample, many of these correlations are likely to reach statistical significance.

For the sake of this discussion, let's assume that we have a random pattern and proceed with one of the methods of imputation suggested earlier. Table 7.7 presents the correlation matrix and scale score statistics for the five-variable analysis using the mean substitution method.

Although the pattern is not perfect, the method of mean substitution generally produces results more representative of the original correlation matrix than does either listwise or pairwise substitution. We present a couple of examples in Table 7.8.

The scale scores, shown in Table 7.9, are also more similar to those obtained from the original data, and the alpha reliability is higher when mean substitution is utilized.

These results indicate that, at least in this example, with 12% of the data points *randomly* assigned as missing, mean substitution would be a reasonable choice for retaining a full data set. In conclusion, no matter what methods are

TABLE 7.8 Comparison of Correlations Based on Different Methods of Imputation/Deletion of Missing Values

Correlation Pair	Item 1/Item 2	Item 2/Item 5	Item 3/Item 5
Original	.3691	.5385	.1550
Listwise	.2571	.3093	.2365
Pairwise	.4417	.5798	.2431
Mean substitution	.3651	.5099	.1644

applied to compensate for missing data, be sure to indicate in any discussion of your research how many values were missing and how they were eliminated or replaced in the various analyses you may have conducted.

HOW DO I CONTROL OR ADJUST FOR OUTLIERS?

Most researchers have been confronted with the dilemma of eyeballing a distribution of data and discovering that a few cases (subjects) have scores that lie far outside the distribution of scores in the sample. Such so-called outliers can be problematic because their presence can unduly affect the description of the sample distribution and subsequent inferential statistics.

In Chapter 2, we discussed how to examine a distribution to detect the presence of outliers using box and whisker plots and stem-and-leaf diagrams. Finding outliers in univariate distributions is relatively simple. An outlier arises as an observation that appears to be unattached to the bulk of the distribution, which is typically piled up near the center, with fewer cases trailing off to the sides. Do outliers contribute to our understanding of the phenomenon being

TABLE 7.9 Scale Scores Based on Different Methods of Imputation/Deletion of Missing Values

Scale Scores	Mean	SD	N	Alpha
Complete matrix	9.69	2.94	49	.68
Missing data matrix	8.11	2.63	21	.58
Mean substitution	9.79	2.69	49	.64

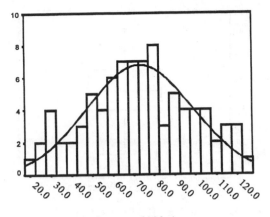

Cups of Water

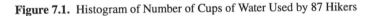

Figure 7.1. Histogram of Number of Cups of Water Used by 87 Hikers

studied? Are the extreme scores from the same population as the other cases in the sample? Should they be kept or deleted in terms of computing statistical summaries and tests?

The first task is to identify the presence of an outlier. Convention suggests that scores that are more than three standard deviations from the mean may be regarded as outliers on a **univariate** distribution. With smaller sample sizes (less than 70), this criterion could be reduced to 2.5 standard scores (z scores); with very large sample sizes, one might anticipate more extreme standard scores (z scores), including a few in excess of three standard deviations from the mean, and adjust the criterion upward. A glance at a frequency distribution or a graphic display can give you a quick indication if an outlier exists. An outlier emerges as a case that appears to be unattached to the bulk of the distribution.

The bar graph in Figure 7.1 illustrates the number of 8-ounce cups of water consumed by 87 hikers on a week-long wilderness expedition. These data represent a continuous distribution with a mean of approximately 73 and a standard deviation of 25.5. According to the three standard deviation rule, an outlier would have a score below $73 - (3 \times 25.5) < 0$ or above $73 + (3 \times 25.5) = 149.5$. As shown by the histogram, and by the boxplot in Figure 7.2, there are no outliers in this distribution.

To understand the impact of an outlier on statistical computations, watch what happens to the mean and standard deviation of the water use data with and without inclusion of outliers. We add to this distribution two cases who con-

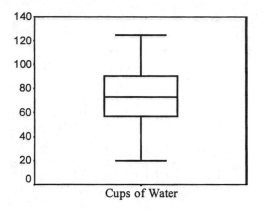

Figure 7.2. Boxplot of Cups of Water Used by 87 Hikers

sumed large quantities of water, 167 and 182 8-ounce bottles each. (The z scores of these two values, in the new distribution containing 89 cases, are 3.11 and 3.62, respectively.) The boxplot, shown in Figure 7.3, clearly shows these values above the upper fence.

The mean of the new distribution is 75.4, up 2.3 bottles from the previous value, and the standard deviation has increased four bottles, from 25.5 to 29.4. Now that we have diagnosed the situation, what do we do about it? Perhaps there is nothing particularly unusual about these data: The two hikers just drink a lot of water. Perhaps the data are in error. For example, one hiker drank 128 bottles, not 182. Although there may be no reason to exclude these two cases from the sample, the researcher might decide to report a different measure of central

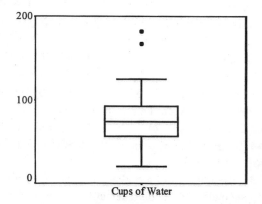

Figure 7.3. Boxplot of Cups of Water Use Among 87 Hikers, With Outliers

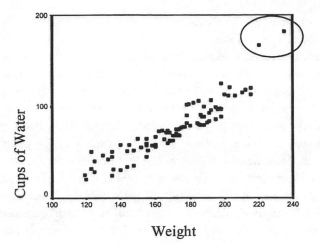

Figure 7.4. Scatterplot of Water Use, by Weight (outliers on use variable circled)

tendency, the median as opposed to the mean, to describe the distribution, because the addition of the two outliers changed the median only from 73 to 74. Our conclusion, based on a comparison of these statistics, probably would be that the outliers were not particularly problematic.

Extreme cases can also exist in a **bivariate** distribution, even when there is no outlier on either single variable that makes up the distribution. Consider the situation where the data indicate that a 17-year-old subject has had three divorces. Three divorces are not uncommon, nor is the presence of a 17-year-old subject, but the combination certainly is unusual. Scatterplots can help visually identify bivariate outliers. Whereas the appropriate measure with univariate distributions is the standard deviation from the mean, with two variables the measure is the **standardized residual** (greater than three) from the regression line. Standardized residuals are commonly computed by most statistical software packages. Superimposing an ellipse that represents a bivariate normal distribution on the scatterplot can help you visually determine the expected range of observations (Hair et al., 1995). The confidence intervals on the ellipse (e.g., 90%) can be adjusted to locate outliers according to whatever criterion you establish. For example, imagine a variable that is likely to be correlated with water use, such as the hiker's weight. We probably would predict a positive relationship between how much a person weighed and how much water was consumed. Figure 7.4 presents the bivariate scatterplot of weight by water consumption. Note that the two outliers do not appear particularly unusual in this representation. We simply happen to have two individuals who weigh a lot and thus drank a lot of water.

Multivariate outliers derived from more than two variables can also be diagnosed, but it is difficult to do so without the aid of statistical software. Stevens (1992) has written a detailed account of how outliers may occur in multiple regression as outliers on the criterion variable or on the predictor variables. The analysis requires determining the multidimensional position of each observation from a common point. The usual measure of multivariate outliers is the **Mahalanobis distance**. The Mahalanobis distance is the distance of a case from the centroid of the rest of the cases, where the centroid is a point in space determined by the means of all the variables (Tabachnick & Fidell, 1996). The Mahalanobis distance is computed using a discriminant function analysis by which an equation is determined that best distinguishes one case from the rest of the cases. Whenever a case has an unusual configuration of scores, those scores become heavily weighted in the function, and the Mahalanobis distance of the case from the bulk of the cases is significant. These computations are available in many statistical software packages, including SPSS REGRESSION®, SYSTAT®, MGLH®, and BMDP.® A conservative p value of .001 or less is recommended to define an outlier using this measure.

Once outliers have been located, there is still the question of what to do with them. Because a primary cause of the presence of outliers is sloppy data recording, the first recommended antidote is to check data entry and transcription for the involved values. Sometimes, missing values are read as real data because missing value codes have not been specified accurately in the computer analysis. In such a case, the correct missing value codes need to be introduced into the analysis. The more complex solutions arise if there are no coding errors. Because it is impossible to know if an outlier is an extreme case within a single population or represents a case drawn from a different population, it is not advisable simply to eliminate it. Outliers easily can be observations that represent a unique but valid aspect of the sample population. These outliers, of course, should be retained in the sample. They contribute to a complete understanding of the phenomenon under study. Elimination of them runs the risk of facilitating the statistical analysis but reducing its generalizability.

If most of the outliers, however, are due to the presence of one variable, it might make sense to delete that variable from the analysis. If the extreme cases were not part of the relevant population of cases, they can be deleted with no loss of generalizability of results because the results do not apply to that population.

If the outliers *do* belong to the sample population, there are two options. One is to retain the cases but modify their values so they won't be overly influential in determining the statistical results of the study. This involves **transforming** the data (Cook & Weisberg, 1982). The transformation is intended to change the distribution of scores to a more normal distribution because the outliers are considered to be part of a non-normal distribution. Transformation allows for easier statistical manipulation and still retains outliers in the tails of the distribution, but with less impact on the results.

A second, and certainly less drastic, option for dealing with outliers is to run your analyses twice, once with the outliers included and once without. Both sets of results can be reported. With reasonable sample sizes, the results from the two analyses frequently will be similar. The point here is that although it is important to examine and diagnose problems or potential problems with your data distributions, sometimes these are much ado about nothing.

HOW DO I ADJUST FOR NON-NORMAL DATA?

We have discussed the normal distribution in a number of places in this book. We first addressed the issue of the normal distribution in Chapter 2 and suggested strategies for assessing the "normality" of one's data. In Chapter 5, we discussed the "assumption of normality" as a basic criterion for the conduct of some statistical tests. In this chapter, we consider the question of what strategies to invoke when one's data do not appear to be normally distributed. This follows from both the material in Chapters 2 and 5 and the handling of missing data and outliers discussed in the previous two sections. Once we deal with missing data and outliers, the problem of non-normal distributions may in fact fix itself; however, when this is not the case, the researcher needs to consider the possibility that some method of adjusting the distribution may be necessary.

Even though many parametric statistical procedures include the requirement of normally distributed *population* distributions, the researcher does not, and cannot, know for certain whether or not the population from which a sample came is normally distributed. He or she can, however, examine the sample distributions for evidence about the population's structure. The larger the sample is, the more confidence we are likely to have in what the sample distributions suggest; small samples are likely to tell us much less. As shown in Chapter 2, the easiest way to get a sense of the shape of a distribution is simply to plot it. The four types of plots that we suggested (histograms, boxplots, stem-and-leaf diagrams, and normal probability plots) are likely to indicate whether or not the distributions contain outliers and/or extreme skew and whether the case for the normality of the population distribution cannot reasonably be made.

Not all distributions are normal. Some would argue that few are, and sometimes extreme cases do belong in the sample, resulting in distributions that are seriously skewed. One option, then, is to modify the distribution in such a way that (a) extreme cases won't be overly influential in determining the statistical results of the study and (b) the distribution assumes a more "normal" shape. This involves "transforming" the data (Cook & Weisberg, 1982).

Transforming data is also useful for responding to a number of distribution problems, such as lack of normality, and in bivariate or multivariate distributions, homoscedasticity, non-linearity, and lack of bivariate and multivariate normality. The biggest limitation of transforming data is the possible difficulty of interpreting the new scores. For instance, if raw scores refer to an inherently meaningful scale such as income, the transformed scores may be harder to interpret. Transforming data nevertheless is generally recommended whenever the assumptions of a statistical test are substantially violated. With regard to outliers, one may transform the data in such a manner that a near-symmetrical distribution is achieved, with fewer outliers, and skewness and kurtosis closer to zero.

Several kinds of transformation are commonly employed. A good introductory reference is Hamilton (1990). More advanced references are Mosteller and Tukey (1977), Box and Cox (1964), and Hamilton (1992). One class of transformations, called **power transformations**, is particularly helpful in reducing skew, condensing outliers, and conditioning the distribution to approximate a normal curve. These are called power transformations because they involve raising the value of a variable (X) to some power (q). For example, squaring a variable raises that variable to the power of 2 (i.e., X^2). (Squaring tends to reduce negative skew.) Tukey (1977) suggested the "ladder of powers," a set of steps that applies different powers to bring non-normal distributions toward normality. We provide some illustrations (Figure 7.5) and an example below.

As can be seen by examining the values of q in Figure 7.5, powers of q that are greater than 1 are used to adjust for problems of negative skew. This is because they tend to change the distribution by shifting the area of the distribution to the upper tail. In contrast, powers less than 1 change the distribution by shifting the area out of the upper tail, thus reducing positive skew. In general, if a distribution differs moderately from normal and is positively skewed, a square root transformation should be tried first, and if a distribution is substantially different from normal, a logarithmic transformation is recommended

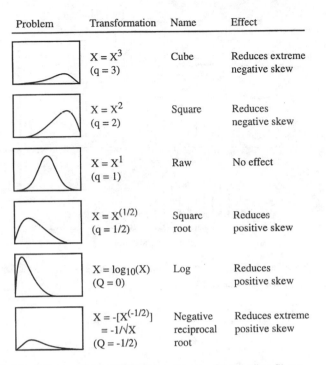

Problem	Transformation	Name	Effect
	$X = X^3$ $(q = 3)$	Cube	Reduces extreme negative skew
	$X = X^2$ $(q = 2)$	Square	Reduces negative skew
	$X = X^1$ $(q = 1)$	Raw	No effect
	$X = X^{(1/2)}$ $(q = 1/2)$	Square root	Reduces positive skew
	$X = \log_{10}(X)$ $(Q = 0)$	Log	Reduces positive skew
	$X = -[X^{(-1/2)}]$ $= -1/\sqrt{X}$ $(Q = -1/2)$	Negative reciprocal root	Reduces extreme positive skew

Figure 7.5. The Effects of Power Transformations on Distribution Shape

(Tabachnick & Fidell, 1996). We provide examples in Figures 7.6 and 7.7 using distributions that represent positive and negative skew.

As can be seen from the examples, a positively skewed distribution can be made approximately normal by applying a log transformation, but we can also go too far and morph a positively skewed distribution into a negatively skewed one, essentially leaving us no better off than when we started. Similarly, "overcorrecting" a negatively skewed distribution by applying a cube transformation changes a negatively skewed distribution into a positively skewed one. A square transformation works much better in this case.

So how does the researcher go about selecting the right power transformation? The answer usually is found by trial and error, but some computer software can help with the job, not only by making it easy to assess the degree of non-normality but also by making suggestions for an appropriate transformation. For example, Stata® (Computing Resource Center) will plot the histogram of a variable using the ladder of powers to let the user select the transformation

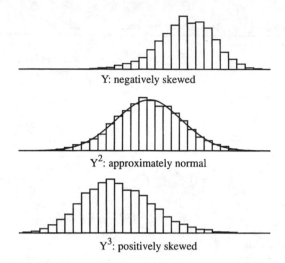

Figure 7.6. The Effect of Square and Cube Transformations on Negative Skew
SOURCE: Adapted from Hamilton (1992).

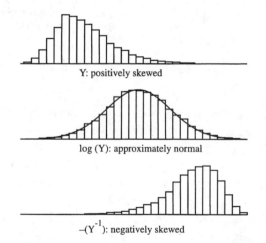

Figure 7.7. The Effect of Log and Negative Reciprocal Transformations on Positive Skew
SOURCE: Adapted from Hamilton (1992).

that appears most reasonable. SPSS® provides a dialog box that contains various transformation options. The user only needs to select one that seems appropriate. Finally, the ladder of powers offers a range of values that may be suitable, but it may be necessary to select a power that is between one of these values. For example, if a transformation overcorrects, such as a cube, this does not imply that a square will be appropriate. It may be necessary to select a power that is between one of the graduated levels we have suggested here, such as a power of 2.3 or 2.5.

Finally, power transformations can also be important in meeting the assumptions of multivariate techniques and can be an aid in dealing with problems of multivariate normality in regression analysis and heteroscedasticity in both analysis of variance and regression analysis. We strongly recommend Lawrence C. Hamilton's intermediate text, *Regression with Graphics* (1992), for an excellent treatment of these applications.

CHAPTER 8

Types of Variables and Their Treatment in Statistical Analysis

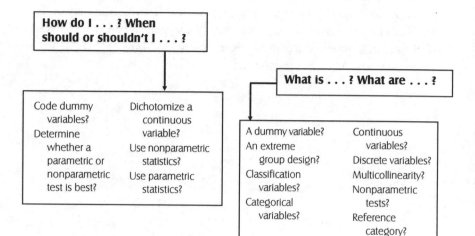

How do I . . . ? When should or shouldn't I . . . ?

Code dummy variables?
Determine whether a parametric or nonparametric test is best?

Dichotomize a continuous variable?
Use nonparametric statistics?
Use parametric statistics?

What is . . . ? What are . . . ?

A dummy variable?
An extreme group design?
Classification variables?
Categorical variables?

Continuous variables?
Discrete variables?
Multicollinearity?
Nonparametric tests?
Reference category?

Level: Intermediate

Focus: Instructional

Types of Variables
and Their Treatment
in Statistical Analysis

The selection of the appropriate method of statistical analysis depends in part on the types of variables that characterize the data. Variables may be categorized as a function of their use in an analysis (e.g., dependent, independent, moderator, mediator) or as a function of their measurement characteristics (e.g., nominal, ordinal, interval, ratio). One of the most useful distinctions, one that we have used throughout this book, is the distinction between **continuous** and **discrete** variables. Discrete variables (also called **categorical** variables) are made up of distinct units in which a finite number of values separates any two points. Both political preference (e.g., Democrat, Republican, Independent, etc.) and religious preference (Protestant, Catholic, Jewish, etc.) are common examples of discrete variables. In contrast, continuous variables reflect a theoretically infinite number of values and can be measured on interval or ratio scales. For example, height and weight both can be considered continuous variables. Often, the same variable can be expressed either categorically or continuously. People can be categorically described as "light," "average," or "heavy," or they can be given a precise weight in pounds and ounces along a

179

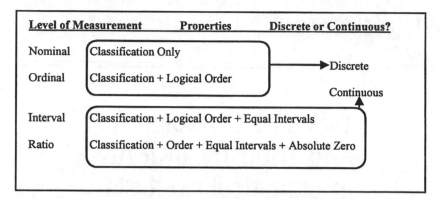

Figure 8.1. Discrete Versus Continuous Variables

continuous scale. Because observed values on any variable are limited by the instruments that measure them, in practical terms any continuous variable could also be expressed as a discrete variable. Figure 8.1 illustrates the distinction between discrete and continuous variables using Stevens's (1966) well-known classification of "levels of measurement" as a guide.

The following sections address three questions that seem to reappear continually on the consultant's desk. All these questions are related to confusion regarding the characteristics of variables and what is gained or lost by treating them in different ways.

HOW DO I DETERMINE WHETHER A PARAMETRIC OR NONPARAMETRIC TEST IS BEST?

Virtually all the discussion of statistical methods throughout this book has concerned what are commonly referred to as "classical" tests. This includes our discussion of ANOVA, MANOVA, regression, and other techniques, all of which are derived from the general linear model. As detailed in Chapter 5, these models are built on a number of assumptions concerning both the measurement properties of the variables and their distributions. To fully meet the assumptions of most of these statistical models, we assume that the variable of interest was sampled from a population with a normal or approximately normal distribution and that the variable is at least intervally scaled. The importance of this is twofold: First, interval scaling permits the use of the mathematical transformations required to calculate these statistics, and second, when distributions are assumed to be normal, only the mean and standard deviation are required to calculate probabilities along these distributions, and, therefore, to reliably set the criterion for the rejection of the null hypothesis. The ascendance of parametric statistical models can be attributed in part to these characteristics.

Numerous authors have pointed out, however, that many psychological, social, and behavioral indicators are simply not normally distributed, or approximate a normal distribution poorly. Sometimes, research studies rely on data that are not derived from interval or ratio scales but represent nominal (categorical) or ordinal (ranking) scales. In such cases, **nonparametric** tests may provide a useful alternative. These tests do not require that the variables be normally distributed, or measured on an interval scale; hence, they are also termed "distribution-free" tests. The chi-square test is probably the best-known nonparametric statistic and requires only the ability to classify cases into a set of categories (i.e., nominal measurement).

We have mentioned previously that changing popularity among statistical tests has been a function, in part, of the increasing availability of technological advances in the manipulation of numbers through the advent of high-speed computers. Nonparametric techniques arose in an era of reliance on hand calculations and were perceived as a suitable alternative to those hand calculations (Kendall, 1962; Siegel, 1956). Many of the tests rely on rank orders rather than raw scores and, as such, are relatively easy to compute. The use of ranks, however, imposes several limitations on the nonparametric framework. First, if there is any information regarding the distance between scores contained in the data, it is lost in the rank transformation. Second, to the extent that the data do represent a normal distribution, the nonparametric model generally will be less powerful than its parametric alternative. The extent of this loss of power is debatable, however. For instance, it has been shown that tests such as the Mann-Whitney U test, the Kruskal-Wallis H test, and the Wilcoxon matched-pairs test, which are all based on rankings, yield high levels of power with only 5% more subjects than the same power with corresponding parametric measures (Jacobson, 1976). The amount of power that a researcher really needs is thus an important consideration.

In many statistics texts, firm prohibitions are made against using a statistic when the level of measurement of the variables involved is inappropriate for that statistic. In fact, some texts use levels of measurement to provide their basic organizational framework. We have no quarrel with the logic of this approach, and we have emphasized the importance of explicit recognition of the underlying measurement model in Chapter 5. The main questions that this chapter addresses are the following: When, in fact, is the more powerful parametric model inappropriate? and, Do we automatically violate the assumptions of parametric models when we utilize an ordinal scale in a multivariate procedure such as multiple regression or factor analysis?

According to Stevens (1966):

> As a matter of fact, most of the scales used widely and effectively by psychologists are ordinal scales. In the strictest propriety, the ordinary statistics involving means and standard deviations *ought not to be used with those scales*, for these statistics imply a knowledge of something more than the relative rank order of data. (p. 26, emphasis added)

The above statement implies that the use of ordinal scales is in fact inappropriate with any technique involving means and standard deviations. Because these statistics are the building blocks of most multivariate techniques, the statement in one broad stroke rules out the use of these techniques with most measures utilized by social scientists.

Unfortunately, as argued by Borgatta and Bohrnstedt (1981), the assumption that "most of the scales used widely and effectively by psychologists are ordinal scales" is not warranted. As they argue:

> Most of the scales used widely and effectively by psychologists (and other social scientists) *simply are not ordinal scales*; the procedures used by social scientists usually fit badly to an interval scale, but they certainly are not ordinal scales. In other words, we measure latent continuous variables with error at the manifest level. (1981, p. 26, emphasis added)

What this means is that the variables of greatest interest to social scientists, such as anxiety, depression, alienation, self-efficacy, and prejudice, are conceptualized (at the latent level) as continuously distributed variables, even though when these variables are measured (at the manifest level) the measurements place the observations into discrete categories. As Borgatta and Bohrnstedt (1981) state, "It is worth emphasizing here *that level of measurement in not a requirement for the use of parametric statistics* as was suggested by Stevens" (p. 28; emphasis added). Furthermore,

> To summarize this argument, most of the central constructs in the social sciences are conceptualized as continuous, and their distributions are such that the appli-

cation of parametric statistics to their analyses will not result in seriously biased estimates. And if the variables are continuous, *they must also by definition, be interval* (1981, p. 29, emphasis added)

Although it is important to acknowledge that our measurements are likely to contain error, and to exercise care in developing these measures so as to maximize their reliability and validity, this does not mean that we must use ordinal level (i.e., nonparametric) statistics to analyze the data. The underlying model with which we approach the analysis of social scientific constructs typically is not one suggesting only operations of "greater than" or "less than" (i.e., ordinality) but rather a model suggesting a continuously distributed variable. Thus, statistical models designed for such variables are not inappropriate.

The theoretical arguments seem to range, on one hand, from the suggestion that using ordinal data to conduct parametric tests violates the mathematical logical system on which these models are based, to the argument that many of these social scientific measurements are in fact interval, not ordinal, and therefore do not violate this system. Statisticians buttress this argument with simulation studies that support the robustness of parametric tests even when the normality assumption is seriously violated (see Zimmerman & Zumbo, 1989). A case in point is the relative insensitivity of the t ratio and the F ratio to violations of normality, an insensitivity that makes them distribution-free tests in practice (Boneau, 1960). There is also evidence to support that scaled data, when ranked, intercorrelate very highly (Stuart, 1954). For example, the correlation between rankings of 25 randomly obtained sets of numbers is about .94 and moves higher with an increasing N. This is similar to comparing a Spearman rank correlation coefficient (a nonparametric measure) and a Pearson product-moment correlation (a parametric measure). To be sure, for several parametric tests Type I error increases beyond that of nonparametric tests for small and/or unequal samples. As Harwell (1988) puts it:

In general, researchers should not automatically rely on the robustness of any parametric test; some are robust and some are not. In no sense can the comparison of the distributional properties of the tests uniformly favor parametric procedures, and researchers who do so commit an error that could seriously affect the statistical results of the analysis. (p. 35)

Finally, Borgatta and Bohrnstedt argue that if we are really considering the data as ordinal, then the computations in any procedure that requires the addition of ranks, or other similar mathematical operation, "make no sense at all" because the mathematical operations are being misapplied. When we convert a set of observations to ranks and assign rank orderings that are then mathematically manipulated, we are in fact assuming an interval scale, because we assume a unit distance between the ranks.

This argument continues with strong proponents on both sides. Unfortunately, most of the multivariate parametric methods do not have an acceptable nonparametric alternative, and we await the development of these tools. The computational power of today's microcomputers, however, has made possible a thorough examination of one's data (see Chapter 2). This at least gives the researcher some basis for making an informed decision regarding the characteristics of the data and their use with parametric methods. Below, we provide some suggestions to assist with this process.

? Recommendations

Our position is that one should attempt to collect data using valid and reliable scales and record that data in a manner that permits retention of the maximum amount of information. Then, *if* our underlying conceptual model describes the variable of interest as continuously distributed, and *if* we have made a reasonable attempt to assess the other necessary assumptions underlying the statistical technique being proposed, there is no reason to reject the use of parametric statistical techniques. We provide a list of specific suggestions below.

1. Make an attempt to select scales that are well validated and have known properties. Avoid the haphazard creation of single-item indicators of concepts.
2. Always examine the distribution of all variables. When distributions are skewed, consider the use of transformations. (See Chapters 2 and 8.)
3. Make use of bivariate plots of data that you plan to use in a single analysis.
4. For seriously non-normal distributions, nonparametric statistics may be appropriate.
5. For truly nominal and ranked data, nonparametric statistics may be appropriate.
6. With small sample sizes and/or unequal groups, nonparametric statistics may be appropriate.
7. If the sample sizes are fairly large (about 60 or more per group) and if group sizes are approximately the same (a size ratio of less than 2:1), parametric methods tend to be robust.
8. When in doubt, use both methods to the extent possible, and if results are the same, report the parametric finding, which generally will be more powerful.

Acknowledge the nonparametric confirmation in your results to corroborate your findings.

9. When the results of parametric and nonparametric testing are different, review the data and select the more appropriate test, stating your reasons for doing so.

? Parametric Tests and Their Nonparametric Alternatives

Typically, when the data are to be considered at least intervally scaled, researchers begin with the assumption that a parametric test will be the statistic of choice, and then, when the data provide an unreasonable fit to the parametric framework, a nonparametric "alternative" is selected. Thus, the situation seems to be one in which the researcher "adjusts" for failure. This approach need not be the case, and we prefer to simply represent parametric and nonparametric tests as options that are appropriate under different conditions. It is not unusual to encounter situations in which the characteristic of the data we wish to examine is clearly addressed by a nonparametric test, and in which a parametric test would be inadequate, inappropriate, or simply not available. For example, Siegel (1956, p. 45) demonstrates the utility of the chi-square goodness-of-fit test for examining the relationship between post position and number of wins in horse racing. (We highly recommend Siegel's classic text, *Nonparametric Statistics for the Behavioral Sciences*, which is more than 30 years old but still impressive in its insights and readability.) Siegel divides tests into six major groupings, as follows: one-sample tests, tests for two related samples, tests for two independent samples, tests for more than two related samples, tests for more than two independent samples, and measures of correlation and their associated significance tests. Thus, in addition to bivariate measures of relationship, the basic distinctions include the number of groups being compared and whether or not group comparisons involve related or independent samples. This distinction is maintained in the recent book by Sheskin (1997), though he adds a category that includes factorial designs and related measures of association/correlation. Kanji (1993) classifies tests as "parametric" or "distribution free" and then lists tests for central tendency, proportion, variability, distribution function, association, and probability within each major classification. Table 8.1 provides a brief summary of some parametric and nonparametric tests. We refer the reader to Siegel (1956), Sheskin (1997), or Kanji (1993) for further test alternatives.

In sum, researchers tend to agree that whenever data are intervally scaled and normally distributed, parametric tests offer more statistical power than nonparametric tests. When such distributions are not normal and sample sizes

TABLE 8.1   Summary of Some Parametric and Nonparametric Statistical Tests and Measures of Association

Type of Test	Parametric	Nonparametric
One sample		
Means	One-sample t test	—
Proportions	—	Binomial test
		Chi-square test
Two sample		
Independent	Two-sample t test	Mann-Whitney U test
		Kolmogorov-Smirnov two-sample test
		Median test
Related	Paired-difference t test	Wilcoxon signed-ranks test
		Walsh test
K-sample		
Independent	ANOVA (F test)	Kruskal-Wallis one-way ANOVA by ranks
		Chi-square test for independent samples
Related	Repeated measures ANOVA (F test)	Friedman two-way ANOVA by ranks
		Cochran's Q
Measures of association	r, eta	Odds ratio
		Log odds
		Lambda
		Kendall's tau
		Spearman's rank correlation

NOTE: For detailed information and examples, see Kanji (1993), Sheskin (1997), and Siegel (1956).

are small, however, nonparametric tests may have an advantage in terms of power as well as the control of Type I error. Distributional violations are more likely to be evident with small sample sizes; hence, nonparametric tests often have been used to analyze such data. Whenever the assumptions underlying parametric tests are met, however, they generally will have more power than the respective nonparametric tests. Moreover, parametric tests such as t tests and F tests are quite insensitive to modest violations of their assumptions. Finally, Cohen (1965) has warned that nonparametric techniques generally are not available for analyzing complex designs. Some aspects of distributions, however, are not addressed by parametric tests. The best advice probably is to use nonparametric statistics to assess distributions in these cases and with nominal

TABLE 8.2 Data Display of Gender and Number of Full-Time Jobs Held

Male		Female	
Subject	*Number of Jobs*	*Subject*	*Number of Jobs*
S1	6	S7	2
S2	5	S8	1
S3	4	S9	5
S4	8	S10	3
S5	7	S11	2
S6	3	S12	3

data, small sample sizes, or significant violations to underlying distributional assumptions. Otherwise, nonparametric statistics have few advantages over parametric choices.

WHAT ARE DUMMY VARIABLES AND HOW DO I CODE THEM?

A **dummy variable** is defined as a dichotomous variable that is usually coded 1 to indicate the presence of an attribute and 0 to indicate the absence of an attribute. Different authors are likely to refer to dummy variables using other terms such as "indicator variable" or "binary variable." The word "dumb" refers to the zero being silent ("dumb") about nonmembership in a category. The advantage of using **dummy coding** of variables is that it allows the use of statistical techniques based on continuous data with variables that are measured as nominal or ordinal (i.e., variables that are discrete). The most common variable that is dummy coded is sex or gender—male or female—but other dichotomous variables, such as "yes" and "no" or "agree" and "disagree," also can be dummy coded.

Let's illustrate how dummy coding might be used to compute a correlation between two variables, one of which is dichotomous. Assume that an investigator wants to examine the relationship between gender and number of full-time jobs held during a lifetime. The data could be arranged as in Table 8.2.

These scores show the number of jobs held by six men and six women in the course of a lifetime. To compute a correlation between number of jobs and gender, however, the data would have to be arranged in a more traditional format.

TABLE 8.3 Data Display of Dummy Coded Gender and Number of Full-Time Jobs

Subject	Gender	Number of Jobs
S1	1	6
S2	1	5
S3	1	4
S4	1	8
S5	1	7
S6	1	3
S7	0	2
S8	0	1
S9	0	5
S10	0	3
S11	0	2
S12	0	3

Assigning "male" the value of 1 and "female" the value of 0 (dummy coding) gives the data set in Table 8.3.

It is now possible to compute a correlation coefficient (**point-biserial** is the appropriate choice) between the variables of gender and number of jobs. In this case, the analysis yields $r = .69$.

In cases where two variables are dichotomous, they can both be dummy coded. Take, for instance, determining the correlation between gender and working outside the home or not. The correlation between two dichotomous variables is called a **phi coefficient**. In this instance as well, the two categories of each variable arbitrarily have been assigned the values of 0 and 1. What we are really doing in these examples is taking a qualitative variable and making it quantitative. We are creating pseudo-scales ranging from 0 to 1 that hypothetically could be labeled according to how we assign the 1 in our dummy coding. For instance, if we give a 1 to the category "Christian" and a 0 to the category "All Others," we have created a code for those who call themselves Christians. Zeros represent those who possess an absence of Christianity, and 1s represent its presence.

A series of dummy variables can be employed when an independent variable has more than two categories. This transformation is often useful when it comes to conducting a regression analysis with discrete variables as predictors. Let's say that one predictor variable is the kind of diet that a subject has been

TABLE 8.4 Data Display of Diet and Physical Fitness

Subject	Diet	Physical Fitness Scale
S1	P	18
S2	P	23
S3	P	15
S4	P	27
S5	C	42
S6	C	33
S7	C	18
S8	C	30
S9	F	26
S10	F	22
S11	F	31
S12	F	26

following: high carbohydrate, high protein, or high fat. The type of diet was recorded by simply supplying a letter to represent the diet category, as shown in Table 8.4 for 12 cases.

One cannot simply assign the numbers 1, 2, and 3 to the three diet categories because that would not create a meaningful scale. Said another way, in line with the previous chapter, we cannot assume that the underlying distribution of diet type is continuous. One could convert diet into three dummy variables, however (high carbohydrate = 1 and other = 0, high protein = 1 and other = 0, and high fat = 1 and other = 0). The data might look like Table 8.5 for predicting the dependent variable of physical fitness.

Note that any group can be described as a function of the other two dummy variables. Thus, in the resulting multiple regression analysis, only two independent variables, carbohydrate and protein, would be entered into the regression equation. This represents one for each degree of freedom ($df = n - 1 = 3 - 1 = 2$). The analysis allows the investigator to evaluate the effects of each newly created dichotomous component. The same technique would allow for a regression analysis of any nominal data with multiple categories. Note also that one cannot have more independent variables in the analysis than there are degrees of freedom because of concerns about **multicollinearity**. In multiple regression analysis, multicollinearity exists when independent variables correlate highly with one another, making it very difficult to evaluate their independent effects

TABLE 8.5 Dummy Coding of Diet With Physical Fitness

Subject	Dummy 1 (D1)	Dummy 2 (D2)	Dummy 3 (D3)	Physical Fitness Scale
S1	1	0	0	18
S2	1	0	0	23
S3	1	0	0	15
S4	1	0	0	27
S5	0	1	0	42
S6	0	1	0	33
S7	0	1	0	18
S8	0	1	0	30
S9	0	0	1	26
S10	0	0	1	22
S11	0	0	1	31
S12	0	0	1	26

on the dependent variable. Said another way, the values of any two of the dummy variables perfectly predict the third, resulting in a multiple correlation of 1.00. Because this represents redundancy, the number of dummy variables entered into a regression equation is always one less than the number of categories of the categorical variable being dummy coded. Sex can be dummy coded with one variable, and a three-category variable, such as diet, can be dummy coded with only two variables.

When using a dummy variable coded in the manner that we have described above, the only statements you can make about the effect of one category of that variable are made in terms of another category. For example, when sex is coded as 0 or 1 with 1 = female, any statements about that variable represent the effect of being female, *as compared to being male.* Similarly, when we utilize the first two of the dummy variables for diet and eliminate the third (high fat), this category becomes what is known as the **reference category**, and when we discuss the effect of diet on fitness, we must speak in terms of high protein diets as compared to high fat diets, or in terms of high carbohydrate diets compared to high fat diets. Thus, when you dummy code as illustrated above, each category is compared to the reference category, which is represented by the zero codes on all other dummy variables. Because the choice of the reference category is

TABLE 8.6 Descriptives: Means and Standard Deviations of Fitness Levels, by Diet Groups

		N	Mean	Std. Deviation	Std. Error	Minimum	Maximum
DIET	Protein	4	20.750	5.315	2.658	15.00	27.00
	Carbo	4	30.750	9.912	4.956	18.00	42.00
	Fat	4	18.750	11.983	5.991	1.00	26.00
	Total	12	23.417	10.184	2.940	1.00	42.00

arbitrary in a statistical sense, you may want to choose a category that would provide a meaningful comparison with the other categories. For example, when dummy coding race (Anglo, Hispanic, African American, Asian), it might make sense to use Anglo as the reference category. Thus, all of the coefficients representing the minority group members would be compared to the Anglo reference category.

The coding scheme for creating dummy variables shown above, referred to as **indicator variable coding**, is not the only method. Another method, called **deviation coding**, allows the researcher to compare the effect of each category of the dummy variable to the average of all categories. We will not discuss the relative merits and arguments surrounding the use of the various methods, but simply point out that there are a number of different methods that might be selected depending on the type of statistical analysis being conducted.

Finally, we show the analysis of the diet and fitness data below. We begin by presenting the descriptive statistics for each diet group in Table 8.6.

The data show that the fitness scores are highest for the high carbohydrate group and lowest for the high fat group. The difference between the high protein mean and high fat mean is $20.75 - 18.75 = 2.00$, and the difference between the high carbohydrate mean and high fat mean is $30.75 - 18.75 = 12.00$. Table 8.7 presents the results of a multiple regression analysis using only the first two dummy variables (D1 and D2), thus making the high fat dummy variable the reference category.

Note that the unstandardized regression coefficients (B) for D1 and D2 represent the difference between the mean fitness score for that category and the reference category (e.g., high fat) and that the constant or slope is equal to the mean of the reference category (18.75).

TABLE 8.7 Regression Coefficients of Dummy Coded Diet Groups, by Fitness Level

Coefficients[a]

Model		B	Std. Error	Beta	t	Sig.
		Unstandardized Coefficients				
1	(Constant)	18.750	4.744		3.952	.003
	D1	2.000	6.709	.097	.298	.772
	D2	12.000	6.709	.580	1.789	.107

a. Dependent Variable: FITNESS

Finally, note that the value of R^2 for this analysis is .289. This is equal to the value of the regression sum of squares divided by the total sum of squares shown in the analysis of variance summary table (Table 8.8).

Would the addition of the third dummy variable have increased the value of R^2? The answer is "no," because the information provided by the third variable is redundant. All the variance explained by a categorical variable can be explained by including one less dummy variable than the number of categories in the categorical variable.

WHEN, IF EVER, SHOULD I DICHOTOMIZE A CONTINUOUS VARIABLE?

Students of statistics generally fall into one of two camps: those who favor the use of analysis of variance techniques and its offshoots (MANOVA, ANCOVA) and those who side with regression techniques (multiple regression and other correlational approaches). Each of these general approaches has its own colorful history and proponents. Lee Cronbach (1957) wrote what has become a classic article on the distinction between them, describing how correlational approaches support research on individual differences, whereas analysis of variance approaches support experimental research. To the experimenter, individual differences constitute within-group error that needs to be minimized. Thus, the ideal experimental group would be one in which everybody was exactly alike, much

TABLE 8.8   ANOVA Summary Table for Regression of Dummy Coded Diet Groups, by Fitness Level

ANOVA[a]

Model		Sum of Squares	df	Mean Square	F	Sig.
1	Regression	330.667	2	165.333	1.836	.214[b]
	Residual	810.250	9	90.028		
	Total	1140.917	11			

a. Dependent Variable: FITNESS
b. Independent Variables: (Constant), D2, D1

like a genetically identical set of laboratory rats. To the student of individual differences, variability is encouraged. Indeed, two variables cannot be meaningfully correlated without a sufficient spread (variance) of scores on the two dimensions.

The choice of method may depend more on the training of the researcher than the research question being addressed. One whose training has been largely in the context of experimental design may think about continuous variables as **classification variables** that are used to classify subjects into groups so that they may become independent variables in an analysis of variance design. For example, sex is a classification variable that automatically produces two groups. Classification variables such as intelligence and extroversion, however, are continuously distributed and do not automatically sort the subjects into groups. Thus, one must divide the variable into groups to make analysis of variance workable. When subjects are selected or sorted into groups on the basis of their scores on an introversion-extroversion scale, the researcher is not conducting a true experiment in that the group differences that become the independent variable are not under the researcher's control. Extroversion is an intrinsic characteristic of the participants, not a characteristic conferred on a random sample of subjects by the researcher through a process of variable manipulation. Thus, the groups may also differ from one another in other unknown ways that are related to the classification variable. The study becomes a correlational study rather than an experimental study. Even if analysis of variance is the chosen inferential statistic, the researcher will have to make do with a statement of association between the variables rather than a definitive conclusion of cause and effect.

One offshoot of the debate about the range of applicability of both ANOVA and regression approaches to data analysis is the issue of when to dichotomize a continuous variable. Does it make sense, for example, to take scores on a variable such as "shyness" or "amount of education" and split the sample in two by comparing subjects who score above the mean on the variable with those who score below the mean? The answer to this question is, "rarely." It is almost never a good idea to throw away information; however, that is precisely what is done by converting a continuous distribution into two or more categories. It is generally much better to conduct correlational analyses than to transform continuous data into independent variables in an ANOVA design. Cohen (1983) has demonstrated that the loss of power that ensues from splitting one variable at the mean is the same as from throwing away 38% of one's subjects! The same argument applies to dichotomizing a variable somewhere above or below the mean, resulting in even greater loss of power. The great majority of studies cannot comfortably afford to reduce power in this way—and there is no reason to.

There are other reasons why dichotomization of a continuous variable typically is a bad idea. One is that the split the researcher makes to create "high" and "low" groups on continuous data may not reflect those distinctions in the real world outside the researcher's particular sample of subjects. For instance, a median split on height among basketball players into "tall" and "short" would be a misleading use of the categories "tall" and "short" from the perspective of the general population. In fact, the dichotomy implicitly assumes that a score near the median is the same as a score at the extremes of the continuum because those distinctions no longer exist.

There are also computational disadvantages to dichotomization. In an ANOVA design with two independent variables, splits on both variables would lead to unequal cell sizes if the predictor variables are correlated (e.g., height and weight). That means that main effects no longer can be added together and that interpretation of the results will be difficult. The same issues are relevant to regression analyses, although multiple regression was designed to cope directly with the implications of correlated predictor variables (Humphreys & Fleishman, 1974).

Some investigators object to using dichotomized individual difference variables in an orthogonal analysis of variance design on theoretical as well as practical grounds (see Humphreys, 1978). Humphreys has argued that measures of individual difference, such as intelligence, are not really independent variables under the experimental control of the researcher, much like the example of introversion-extroversion cited earlier in this section. They are invariably correlated with other individual difference variables and are thus not truly orthogonal to each other. Imposing orthogonality on non-orthogonal variables leads to distorted relationships between them and the dependent variable. Moreover, the representation of the variables in the contrived sample groups is not likely to be proportional to their representation in the population. We agree with Humphreys's concerns; however, we note that in current practice pseudo-orthogonal designs are very common and the term "independent variable" is used quite liberally with individual difference variables and within correlational models. The key is to exercise care and restraint in interpreting the results of statistical analyses so that interpretations are a logical extension of the form of the data rather than the machinations of the statistical procedure.

Finally, note that it is typically inappropriate to categorize the *dependent* variable and proceed with traditional regression analysis. Despite the fact that this is exactly what discriminant function analysis does, when the restrictive assumptions of this procedure are not met and/or when the independent variables are also dichotomous, it is possible to obtain predicted values from the discriminant function equation that are biased (see Hosmer & Lemeshow, 1989, pp. 18-20). There are other techniques that can deal effectively with dichotomous dependent variables. These include logistic regression and logit analysis. When the dependent variable is continuously distributed, leave it that way.

? Recommendations

Now that we have dissuaded you from joining a commonly accepted practice in statistical analysis, it is important to note that there still is a place for the dichotomization of continuous data in research design. That place is prior to the analysis of the data, when the study is being designed. The so-called **extreme groups design** is an excellent way to increase power of a study at the outset and is especially appropriate for studying weak or moderate treatment effects. This strategy implies selecting for comparison only those subjects who score at the extreme ends of a distribution on an independent variable. A commonly employed version of this design is for the investigator to compare subjects who score in the top third of a distribution with those who score in the bottom third on a pertinent dependent variable. Such a design will have more power than comparing the top 50% of a sample with the bottom 50% because the more ambiguous middle range of scores has been removed. The same argument can be extended to using even greater extremes of the distribution, such as the top 10% compared to the bottom 10%. If psychotherapy has an impact on well-being, for instance, it stands to reason that a study that compares subjects who have had high doses of psychotherapy with those who have had virtually no psychotherapy will be more powerful than a study that compares subjects who have had more therapy with those with less therapy, including whatever distinction exists at the mean or median of the distribution. (This argument assumes, of course, that there is a linear relationship between the independent and dependent variables. Moreover, there may be other merits in evaluating a dose effect in psychotherapy.)

Although the advantages of increasing power in this way might seem self-evident when designing a study, note that there are disadvantages as well. The key disadvantage is that extreme scores on a distribution may not be representative of the population or of the phenomenon being studied. For example, a researcher might screen 1,000 subjects on a variable such as aggressiveness and then retain the 100 most aggressive and the 100 least aggressive for the study. A reduction in Type II error (greater power) might be offset by an increase in Type I error, however, because the results may not be replicated easily within the same population.

Another situation in which the dichotomization of a continuous variable might be appropriate is related to the concept of "clinical cutpoints." Most measures of psychopathology that have been "normed" suggest values or cutpoints above which a subject can be considered "clinical" or in need of clinical intervention or some other form of treatment. If the stated purpose of the research is to identify such clinical cases and compare them to "nonclinical" cases, the dichotomization of the variable in question at the clinical cutpoint may be justified.

PART four

Understanding
the Big Two

Major Questions About Analysis
of Variance and Regression Analysis

In the first three parts of this book, we have described and confronted issues that relate to the general understanding and practice of contemporary statistics in the social sciences. In this final section, we move directly into issues and controversies that pertain to the two most prominent families of statistical approaches: analysis of variance and multiple regression analysis. These two techniques, both variations of the general linear model, generate the majority of questions that we typically receive from students and colleagues. The last chapter offers some thoughts on weaving together the separate strands of the book into a more integrated pattern.

Chapter 9 addresses several questions relevant to using the analysis of variance technique. The chapter begins with a nuts and bolts description of this popular method and is designed for the reader who needs a brief refresher course. Thereafter, the content of the chapter becomes increasingly advanced. The first question responds to confusion about how to understand an interaction effect and includes a working example. The next question explores pretest-posttest designs, which usually are employed to measure change based on experimental interventions. A number of statistical options are available for the enterprising researcher. That is also true in the answer to the next question, which inquires about the difference between planned and post hoc comparisons. Both of these topics are pitched at an intermediate level of understanding. In the following section, we describe the "familywise" error rate and how to control for it. This is an important, and somewhat controversial, issue and demands a more sophisticated level of understanding of statistics. The final question in this chapter asks about the appropriate use of multivariate, as opposed to univariate, analysis of variance. This discussion presupposes a basic understanding of ANOVA (analysis of variance).

Chapter 10 considers issues that are relevant to the second major variation of the general linear model, namely, regression analysis. This approach to data analysis seems to be increasingly prominent in the social sciences, especially in response to research questions that are not strictly experimental in nature. The chapter begins with a brief description of the method, intended for those who are uninitiated or rusty in their practice of it. Thereafter, the discussion moves from an intermediate to a more advanced level. The first section asks the pragmatic question of how many predictor variables to use and how many subjects are needed. The next section explains the different kinds of regression analysis that exist and the appropriate circumstances for employing them. The next question asks how to interpret regression coefficients. This is part of the sometimes confusing task of how to make sense of regression output from computer analyses. The final question in this chapter is a return to the topic of interaction effects, in this case in the context of multiple regression analysis.

Chapter 11 is our attempt to paint the issues we have introduced in this book on a larger canvas. Statistics is usually taught in separate units, so that many students are never introduced to the relationship among various statistical techniques. Mathematics is a tool that can be used to translate one type of statistical technique into another. In the first section of this final chapter, we offer one approach to establishing a seamless connection among different techniques. This description should be clarifying for those readers with a little background understanding of the separate methods. The second section of the chapter introduces meta-analysis, a rapidly proliferating method of analysis that is consistent with the epistemological changes occurring in contemporary statistics that are described in Chapter 4. This necessarily brief introduction to the topic will take the uninitiated reader only so far but can be used as an entry point for further study. The final section of the chapter is a recapitulation of some of the themes that have been addressed throughout the book. Here we offer, in summary form, 10 suggestions for maximizing the likelihood of success for your statistical analyses.

CHAPTER nine

Questions About Analysis of Variance

What is . . . ? What are . . . ?

How do I . . . ? When should or shouldn't I . . . ?

Assess substantive significance?

Calculate a reliable change index?

Calculate contrast weights?

Choose between MANOVA and ANOVA?

Conduct a contrast analysis?

Conduct post hoc comparisons?

Construct a classification plot?

Construct an ANOVA summary table?

Control for familywise error rate?

Control for familywise error rate?

Interpret an interaction effect?

Interpret analysis of covariance?

Measure substantive change as opposed to statistically significant change?

Select the appropriate analysis of variance design?

Select the best method for analyzing the pretest posttest design?

Select the best test for post hoc comparisons?

Use ANCOVA?

Use ANOVA?

Use MANOVA?

A classification plot?

A planned comparison?

A time-series design?

An analysis of variance?

An F ratio?

An interaction effect?

An omnibus test?

An STP?

Analysis of covariance?

ANOVA, ANCOVA?

An ANOVA summary table?

A posteriori tests?

A priori comparisons?

Bonferroni correction?

Change scores?

Contrast weights?

Contrasts?

Covariate?

Eta squared?

Gain scores?

Homogeneity of regression?

MANOVA, MANCOVA?

Methods for measuring change?

Multiple comparison methods that are available?

Multiple comparisons?

Multiple comparison tests?

NHST?

Post hoc comparison?

Qualitative interaction?

Quantitative interaction?

Regression to the mean?

Reliable change index?

Statistical interaction?

The difference between planned and post hoc comparisons?

The difference between redundant and orthogonal contrast weights?

The difference between within and between

Level: Intermediate/Advanced

Focus: Conceptual/Instructional

Questions About
Analysis of Variance

●∿ For many decades, the **analysis of variance** has been regarded as the primary method for evaluating the results of experimental studies in the social sciences. The status attributed to analysis of variance is a direct legacy of the seminal contributions of Sir Ronald Fisher to the statistical comparison of treatments early in the 20th century. Fisher's work in statistics began in the area of agricultural research at the Rothamsted Experimental Station in England in 1919. Prior to Fisher's arrival, agricultural experiments at Rothamsted consisted of dividing a field into several *plots*, each of which would receive a different manure or fertilizer treatment in an effort to maximize crop yields (Cowles, 1989). It was evident, however, that differences in yield might be due to factors other than fertilizer, such as differences in soil quality throughout the field. Fisher's experimental design consisted of dividing a field into *blocks*, then subdividing each block into plots, giving each plot within the block a different, randomly assigned, treatment. This design formed the basis for the method of analysis known as the analysis of variance (ANOVA), described for the first time in a paper published in 1921 and used two years later in a published experiment evaluating the effect of manure on potatoes (Fisher & MacKenzie, 1923).

At the heart of Fisher's analysis lay two principles (Cowles, 1989). The first is the key role played by deliberately induced **randomization** in controlled experimentation. This led to the possibility of eliminating any systematic variability not associated with the chosen treatment. It also ensured the derivation of a valid estimate of error, which relates directly to the second principle, the possibility of controlling random error. By replicating each treatment within blocks, random error could be attributed to variability among plots in the same block, and variability between blocks could be removed. Fisher's analysis of variance procedures and his explanation for the crucial role of randomization in designing experiments are fully described in *Statistical Methods for Research Workers*, initially published in 1925, as well as in *The Design of Experiments*, published in 1935.

WHAT ARE THE NUTS AND BOLTS OF ANALYSIS OF VARIANCE?

Analysis of variance is used for assessing the statistical significance of the relationship between categorical independent variables and a single continuous dependent variable. It evaluates experimental hypotheses by assessing treatment effects by comparing the means between two or more groups of subjects that are treated differently. It is assumed that differences **between** the scores of the groups (e.g., differences in group means) will be due to a combination of a systematic treatment effect and unsystematic group differences (random error). It is also assumed that differences in the scores **within** each group will be due to unsystematic individual differences (random error).

This is quite a bit of information for one short paragraph, so let's provide a simple example. Imagine a study of three groups of five subjects each. Two groups each receive a different treatment, and one group is a control group. For example, imagine we are comparing the effect of two different exercise programs on heart rate following the 100-meter dash. The control group receives no training. The setup of this design would look like Table 9.1.

The X values in Table 9.1 represent scores for each case, and the subscripts represent the case number and group number; thus, X_{32} represents the third observation in the second group. Now let's visualize the above 15 cases in a graph that shows where each case lies.

What the graph in Figure 9.1 illustrates is that the mean scores for treatment group 1 are lower than those for group 2, and both of these are lower than for the control group. Generally, the individual scores also tend to be lower in group 1 than in group 2, and both differ from the control; however, this is not always the case. In more common language, we would like to attribute the differences

TABLE 9.1   A Three-Group Experimental Design

Treatment 1	Treatment 2	Control
X_{11}	X_{12}	X_{13}
X_{21}	X_{22}	X_{23}
X_{31}	X_{32}	X_{33}
X_{41}	X_{42}	X_{43}
X_{51}	X_{52}	X_{53}

between the group means to the effects of the treatment. This is considered differences (or variability) *between groups*. These differences, however, also could be due to random error. Thus, the between group differences (variability) could be attributable to both treatment differences and random fluctuations in the value of the group mean.

We attribute differences between the cases *within a group* to factors other than treatment differences. Differences between cases within a group cannot be attributed to treatment differences because the cases all received the same treatment. We call this *within group variability* and attribute it to other factors that randomly affect the individuals in the study. This is considered the random error component.

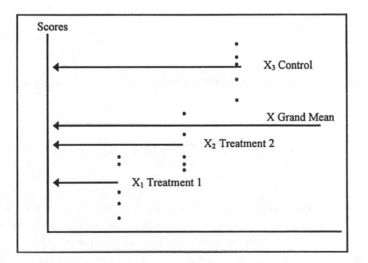

Figure 9.1. Plot of a Three-Group Experimental Design

Our statistical question is whether or not the group means are "significantly different." This means that we must rule out the possibility that the between group variability (i.e., the differences between the group means) is due solely to random error. To do this, we start with the assumption that both the between group and the within group differences are due totally to random error. This is the null hypothesis in analysis of variance.

From these assumptions, a useful ratio is formulated:

$$\frac{\text{Differences Among Treatment Means}}{\text{Differences Among Subjects Treated the Same}}.$$

Shown another way:

$$\frac{\text{Between Group Differences}}{\text{Within Group Differences}}.$$

If we assume that there is a treatment effect (i.e., mean differences are due to the treatment), this effect can only express itself in the numerator of the above ratio. This is because only the numerator contains **between** group differences. With a treatment effect, this ratio becomes

$$\frac{\text{Treatment Effects} + \text{Error}}{\text{Error}}.$$

Under the null hypothesis, suggesting *no treatment effects*, there is no reason to believe that the differences between treatment means should be any different from the differences among subjects within a group (i.e., treated the same). Thus, the value of the above ratio should be 1 or close to 1. With no treatment effect, this ratio approaches the value of 1, because the ratio contains only two estimates of error, as shown below:

$$\frac{\text{error}}{\text{error}} \cong 1.$$

The stronger the treatment effect is, the larger the ratio becomes, because the numerator is increased relative to the denominator as the treatment effect increases (i.e., the mean differences, or effect size, becomes larger). The test statistic for the analysis of variance is called F. The F ratio compares the average variability between group means with the average variability of scores within groups. These differences represent variability and are expressed as "variances" because of their computational advantages. We compare the variance between means of the treatment groups to the variance within groups, created by comparing each score within a group to its own group mean, to determine how large the mean differences are on average. The F statistic, in its most simple incarnation, becomes

$$F = \frac{Between\,Group\,Variability}{Within\,Group\,Variability}.$$

Note that within group variability, representing random error, may also be referred to as the error or residual term or, simply, unexplained variability. There is very little consistency between computer programs as to how this term is labeled. In addition, as analysis of variance designs become very complex, the exact nature of the appropriate error term may be difficult to determine.

The importance of the F ratio lies in the fact that the probability of any obtained value of F is known (under the assumption that there are no treatment effects [i.e., the "null hypothesis" or H_0 is true]). When the value of F is large enough to be unlikely given the assumption of no treatment effects, we reject the assumption and conclude that treatment effects must be influencing the size of the F ratio. Said another way, we conclude that the findings are "statistically significant."

The analysis of variance divides up the total variance of all the scores into separate sources of variance that can be compared for testing for statistical significance. This is called the "partitioning" of the variance. For the one-way analysis of variance model we have shown above, the total variability would be partitioned as follows:

$$\Sigma(X_{IJ} - \overline{X})^2 = \quad \Sigma(X_{IJ} - \overline{X}_J)^2 \quad + \quad \Sigma n(\overline{X}_J - \overline{X})^2$$
$$\text{Total SS} = \text{Within Group SS} + \text{Between Group SS}$$

Where:

X_{IJ} = a score (i.e., score of case I from group J)

X_J = a group mean (i.e., mean of group J)

$\overline{X}$ = the grand mean (i.e., mean of all scores)

n = number of cases in a group

SS = sum of squares

Note that the above equation shows the total sum of squares on the left. This is the sum of the squares of the difference of each score from the grand mean and is the numerator of the equation for the variance. This total is divided (partitioned) into two parts: the difference between the individual scores within a group and the group mean (summed across all groups), and the differences between the group means and the grand mean (multiplied by the number of cases in each group and summed across all groups). The first component on the right represents the within group variability or within group sum of squares, and the second component on the right represents the between group variability or between group sum of squares. Thus, in a "one-way" analysis of variance the variability is partitioned into only two pieces, one labeled **between group** variability and the other labeled **within group** variability. In more complex designs, the variability is partitioned into portions that are due to the individual factors (i.e., the independent variables or main effects) and the interaction of these variables (i.e., the interaction effects).

The analysis of variance can also provide an index of effect size, called eta^2, that measures the proportion of the variance in the scores accounted for by the experimental treatment. **Eta^2** is defined as the ratio of the between group sum of squares to the sum of the between and within group sum of squares:

$$Eta^2 = \frac{SS_{between}}{SS_{between} + SS_{within}}.$$

Examination of the above equation shows that we have taken the between group sum of squares and divided it by the sum of the between and within group sums of squares, which is in fact the total variability (total SS). When there is no error, SS_{within} will be zero and the above equation will yield $eta^2 = 1$, indicating that 100% of the variability is explained by the treatment or between group effect. When there are no treatment effects, the $SS_{between}$ will approach zero, and the effect size will tend toward zero. Because the sum of the between and within groups sum of squares (SS) is equal to the total sum of squares, eta^2 is essentially equal to the ratio of the between group to the total sum of squares and represents the proportion of explained variability. (In regression analysis, variability is partitioned into a total sum of squares and a sum of squares due to regression. When the latter is divided into the former, R^2 is produced. This is directly analogous to eta^2.)

Finally, we will end our introduction to the logic of analysis of variance with two tables. The first table outlines the basic structure of what is known as the

TABLE 9.2 The ANOVA Summary Table

Source	Degrees of Freedom (df)	Sum of Squares (SS)	Mean Squares (MS)	F
Between groups	$p-1$	$SS_{between}$	$SS_{between}/df$	
				$MS_{between}/MS_{within}$
Within groups	$N-p$	SS_{within}	SS_{within}/df	
Total	$N-1$	SS_{total}		

NOTE: F = observed value of F ratio, p = number of groups, and N= number of subjects.
Hypotheses: Null is H_0: $\mu_1 = \mu_2 = \mu_3$; alternative is H_1: $\mu_i \neq \mu_j$ for some i and j.
Assumptions: Subjects are randomly sampled, groups and subjects are independent, and population variances are equal.

analysis of variance summary table, or just the ANOVA summary table. The format of this table is highly consistent in statistics textbooks, computer programs, and published articles. The second table outlines the "flavors" of analysis of variance and should serve as another guide in helping you select the appropriate type of ANOVA to conduct.

Table 9.2 shows that one divides the between groups sum of squares by its degrees of freedom to obtain the between groups mean square and the within group sum of squares by its degrees of freedom to obtain the within groups mean square. Then the between group mean square is divided by the within group mean square to obtain the value of F. This value is then compared to a table of F values to determine its statistical significance. More than likely, we would obtain the exact probability of this result under the null hypothesis from the statistical program package we were using.

Table 9.3 shows a number of different types of analysis of variance designs. When we move from one independent variable to more than one, we change from one-way ANOVA to factorial ANOVA, and the number of effects to be evaluated increases rapidly. With two independent variables, there are three effects to be evaluated, including one interaction. With three independent variables, there are seven effects to be evaluated, including four interactions. The implications of this will be explored in the sections that follow.

The issues surrounding the analysis of variance are many and fill dozens of texts and hundreds of articles. In our experience, a relatively finite number of questions continue to frustrate those who utilize ANOVA designs in their research. These questions center around the nature of interaction effects and their

TABLE 9.3 Selecting the Appropriate Method for Analysis of Variance Designs

Number of Independent Variables	Number of Categories of Each Independent Variable	Type of Design	Type of Test	Effects
1	2	Two group (2 means)	t test or one-way ANOVA	1: Between groups
1	3+	Multigroup (3+ means)	One-way ANOVA	1: Between groups
2	2	2 × 2 factorial (4 means)	Factorial ANOVA	3: 2 main, 1 interaction
3	3, 2, 2	3 × 2 × 2 factorial (12 means)	Factorial ANOVA	7: 3 main, 3 two-way, 1 three-way interaction
4	3, 3, 4, 2	3 × 3 × 4 × 2 factorial (72 means)	Factorial ANOVA	15: 4 main, 6 two-way, 4 three-way, 1 four-way interaction

NOTE: All these designs assume one dependent variable, continuously distributed.

interpretation, the analysis of the pretest-posttest design, the nature of planned and post hoc comparisons, the concept of experimentwise error rates, and, finally, the use of MANOVA. We address each of these questions in the six sections that follow.

WHAT IS AN INTERACTION EFFECT AND HOW DO I INTERPRET IT?

An **interaction effect** is the joint effect of two or more independent variables on a dependent variable; however, two variables can both influence a dependent variable without an interaction being present. In the relatively simple two-variable case, an interaction effect exists if you cannot accurately predict the effect of one variable on the dependent variable without knowing the value of the second variable. Said another way, an interaction occurs when the effect of one independent variable is dependent on the level of another independent variable

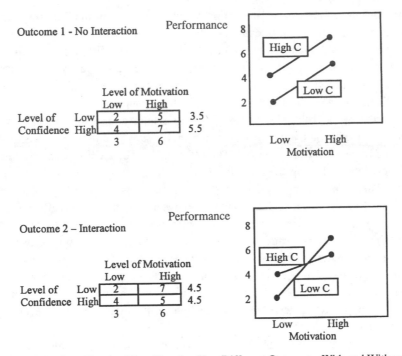

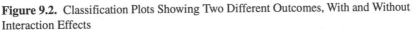

Figure 9.2. Classification Plots Showing Two Different Outcomes, With and Without Interaction Effects

or is moderated by that variable. Single-factor designs (i.e., one-way ANOVA) allow only for the possibility of simple effects; factorial designs allow for the possibility of both main effects and interaction effects. A simple example will make this clear.

Consider a 2×2 factorial design examining the effect of level of motivation (high, low) and level of confidence (high, low) on academic performance. Figure 9.2 presents two examples of potential results on a 10-item multiple choice test. We assume that the cell sizes are equal and that the tables show the cell and marginal mean scores.

In the first example, we can see that both motivation and confidence have a positive effect on performance. Highly motivated persons score three points higher, regardless of which level of confidence they represent. The same is true for confidence: Highly confident people score two points higher, regardless of how motivated they are. The graph to the right of the data table shows the classification plot representing this outcome. The lines are parallel, which indicates the absence of an interaction effect. The slopes of the lines represent the effects of motivation. Because the slopes are identical, the effect of motivation is identical, regardless of confidence grouping. The distance between the lines represents the effect of confidence. The distance is exactly two points, regardless of motivation group. Because the distance is identical, the effect of confidence is the same, regardless of motivation grouping.

In the second data table, the average effect of motivation is the same: A move from low to high motivation results in an overall change of three points, as shown by the marginal means (3 vs. 6). The effect of confidence has been changed, however, by switching the confidence effects for the high motivation groups only. There is now no main effect of confidence, as shown by the marginal means (4.5 vs. 4.5). We now must interpret the findings differently. Specifically, for low levels of motivation, the effect of high confidence is to increase the scores, but for high levels of motivation, the effect of high confidence is to decrease the scores. Said another way, if your confidence is low, the effect of high motivation will be greater (five points) than if your confidence is high (one point). The classification plot graphically displays this result. The line for the low confidence group is quite steep, whereas the line for the high confidence group is rather flat.

Although the above example illustrates the basic ideas underlying the notion of an interaction effect, the actual interpretation of interactions can be much more difficult, particularly when more than two independent variables are involved. One can also have a second-order interaction when three variables interact, a third-order interaction when four variables interact, and so on. In actual practice, one seldom attempts to interpret more than a second-order (i.e., three-way) interaction.

Generally speaking, the only requirement for a nonzero interaction effect in a classification plot, when graphed, is that the lines are not parallel. The lines do not have to cross or form an "X" for an interaction to occur. (Of course, non-parallel lines will always cross at some point.) For example, it may be that fear-based and information-based anti-smoking campaigns are equally effective (or non-effective!), but each strategy will work only with certain groups of people. Perhaps older smokers will be influenced to reduce cigarette consumption only through a rational presentation, whereas younger smokers will be discouraged from smoking only when sufficiently frightened about the long-term effects. The graph in Figure 9.3 suggests that there is no main effect for either presentation strategy or age, but there is an interaction effect.

Treatment × context interactions (in the above example, intervention × age) are very common in two-way ANOVAs. In fact, William McGuire (1989) has

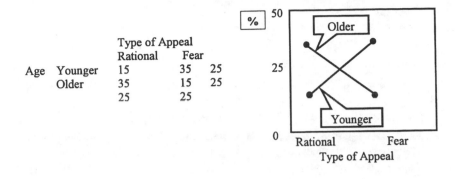

Figure 9.3. Relationship Between Type of Anti-Smoking Appeal, Age, and Percentage Reduction in Smoking

suggested that one of the more creative ways for developing new hypotheses is to take a well-established relationship and imagine circumstances under which its opposite would hold. These circumstances, in the form of a contextual variable such as age, can either increase, reduce, or reverse a treatment effect. One distinction cited in the literature is between **quantitative** and **qualitative** interactions (Schaffer, 1991). A quantitative interaction, sometimes also called an ordinal interaction, refers to two simple effects that are in the same direction but differ in strength, yielding a significant main effect and a significant interaction. A qualitative interaction, sometimes also called a disordinal inter-action, as in the above example, is a function of simple effects in opposite directions and warrants a stronger emphasis in the reporting of results.

So, what is the problem? The problem is that although the definition of a statistical interaction is clear and unequivocal, the *interpretation* of interactions is not at all clear. In an eye-opening survey of 191 research articles relying on ANOVA designs involving interaction, Rosnow and Rosenthal (1989a) found that only 1% of the articles interpreted the interactions in a totally correct manner! There are basically three components to the interpretation of an interaction effect. The first is to determine whether or not an interaction effect actually exists in the population. If we determine that an interaction actually exists, the second question to address concerns the strength of the effect, usually expressed in terms of some measure of effect size or variance explained. Finally, we must address the exact nature of the interaction in terms of both its empirical understanding and its theoretical relevance.

? What Are the Issues in the Interpretation of Interaction?

Rosnow and Rosenthal (1989a) have argued that the source of the error in interpreting interactions is very consistent. When researchers identify significant interactions, they look at the differences among the original cell means to interpret them. But these cell means, which represent simple effects, are composed of main effects as well as interaction effects. Although the crossed lines generated by group means in two-way designs can imply the presence of an interaction effect, these values themselves are not the interaction. Mathematically, interactions are defined as *residual* effects that are left over after other effects contributing to the group means have been removed. In a 2×2 table, this means removing main effects for each independent variable. With higher order designs, such as three-way ANOVAs, it means removing the three two-way interactions as well as the three main effects.

There appears to be a lack of congruity between the mathematical meaning of interaction and the meaning of the term in common usage. Somehow, the layperson's use of the term has contributed to a foggy understanding of the term in a scientific context. Moreover, there is some controversy regarding whether or not interaction terms obscure or contribute to a proper understanding of scientific phenomena. Lack of clarity is especially common in trying to usefully interpret higher order interactions beyond the two-variable case. Computers may have no trouble generating analyses of such interactions, but they are difficult to describe visually and to understand conceptually. A researcher might be prudently advised to interpret interactions for the two-variable, and perhaps three-variable, case, but to structure studies with multiple variables to focus on theoretically meaningful relationships rather than generating all possible interactions among variables.

A final issue relating to the interpretation of interaction terms has been eloquently raised by Abelson (1995) in a recent book. Abelson notes that most two-way ANOVAs necessarily use fixed levels of the two factors, which may lead to serious limitations in generalizing the results of the statistical analyses to anything beyond the specific levels chosen by the researcher. Drawing once more on the example of anti-smoking treatments and age of the smoker, a disadvantage of comparing 18-year-old smokers with 40-year-old smokers is that one cannot rightfully imply that the nature of the interaction is consistent or generalizable across any other age values that might be compared.

TABLE 9.4 ANOVA Summary Table for the Effects of Motivation and Confidence on Academic Performance

Source	df	Sum of Squares (SS)	Mean Squares (MS)	F
Motivation (A)	1	75	75	21*
Confidence (B)	1	50	50	14*
A × B	1	25	25	7*
Within groups	98	350	3.57	
Total	99	500		

NOTE: $F_{.01}(1, 80) = 6.96$.
*$p < .01$.

? Recommendations

In the introduction to this chapter, we suggested three components to the understanding of interaction effects: detecting their existence, assessing their strength, and interpreting their meaning. The first two of these can be accomplished by reference to the analysis of variance summary table. Take, for example, our study of motivation and confidence on academic performance. Table 9.4 presents the results (contrived) of this study in an ANOVA summary table with 100 subjects, 25 per group.

The results of this study show that all three effects are statistically significant at $p < .01$, thus supporting the conclusion that an interaction effect exists in the population (within the limits of Type I error). Note, however, that the sums of squares for the main and interaction effects differ. We can thus partition the total variability, represented by the total sum of squares, into four components: two main effects, one interaction effect, and error (unexplained, residual, or within groups variability). The proportion of this variability explained by the interaction term can thus be calculated by dividing the total sum of squares (500) into the sum of squares resulting from the A × B interaction (25). This equals .05 or 5%, indicating that 5% of the variability in performance is due to the interaction of confidence with motivation. This is equivalent to the eta^2 value for this effect.

TABLE 9.5 Attention Level Means, by Gender and Method of Discipline			
	Boys	*Girls*	*Mean*
Method 1	50	40	45
Method 2	30	60	45
Mean	40	50	45 (Grand Mean)

Finally, we come to the issue of interpreting the interaction effect. Rosnow and Rosenthal (1989a, 1995) have contributed much to our understanding of the proper interpretation of interaction effects. The following discussion, using original examples, is based on their recommendations. The overall strategy is to look at cell means that have been corrected by removing main effects. This strategy is relevant whenever you conduct omnibus F tests that yield main and interaction effects. Whenever planned contrasts are employed (see Chapter 4), it may make a great deal of sense to compare original cell means. We also strongly suggest that one always use classification plots to assist with the interpretation of interaction effects.

Let us assume that an investigator has used two forms of classroom discipline to control the behavior of male and female grade school students. The dependent variable is a measure of paying attention to the teacher. The 2×2 table of results looks like Table 9.5.

The data, as presented, yield the classification plot in Figure 9.4, which suggests that an interaction effect exists. Although there is no main effect of method, method 1 seems to work better with boys and method 2 seems to work very poorly with boys and much better with girls.

To get a more complete understanding of the nature of the interaction, the main effects need to be subtracted from the cell means. First, the row effects are removed from each row by subtracting the mean of each row from the mean of the total sample. The table is now corrected for the row effects (see Table 9.6).

Next, the column effects are removed from each column by subtracting the mean of each column from the grand mean (see Table 9.7).

Table 9.7 is now corrected for both row and column effects, showing no main (row) effect of method and a small main (column) effect of gender.

Finally, the grand mean is removed from the overall effects, represented as the cell means. This step is advised because the grand mean adds a constant value to the residual values. By removing this constant, it is easier to display and interpret the results. We show this modified table as Table 9.8.

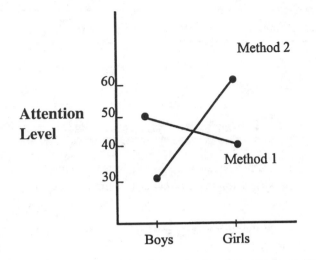

Figure 9.4. Classification Plot of Attention Level, by Gender and Method of Discipline

TABLE 9.6 Attention Level Means, by Gender and Method of Discipline, Corrected for Row Effects

	Boys	*Girls*	*Mean*	*Row Effect*
Method 1	50	40	45	0
Method 2	30	60	45	0
Mean	40	50	45	

TABLE 9.7 Attention Level Means, by Gender and Method of Discipline, Corrected for Row and Column Effects

	Boys	*Girls*	*Mean*	*Row Effect*
Method 1	50	40	45	0
Method 2	30	60	45	0
Mean	40	50	45	
Column effect	−5	+5		

TABLE 9.8 Table of Adjusted Cell Means and Row and Column Effects

	Boys	Girls	Row Effect
Method 1	+5	−5	0
Method 2	−15	+15	0
Column effect	−5	+5	

If there is no interaction effect, we should now be able to reproduce the values of the adjusted cell means by summing the row and column effects. We show this in Table 9.9.

TABLE 9.9 Table of Predicted Cell Means

Cell	Row Effect +	Column Effect =	Predicted Cell Mean
1, 1	0	−5	−5
1, 2	0	5	5
2, 1	0	−5	−5
2, 2	0	5	5

Clearly, these values do not correspond to the adjusted cell means in Table 9.8. Thus, we can calculate the "residual" value by subtracting the estimate from the adjusted cell means, as in Table 9.10.

The resultant table of residuals (residual = group mean − estimate) reveals an interaction, signified by the X-shaped pattern shown in Table 9.11. The interaction is the difference between the means of the two diagonals and will yield an X-shaped plot (although other patterns are possible with more levels of a variable).

TABLE 9.10 Table of Interaction Residuals

Cell	Cell Mean −	Estimate =	Residual
1, 1	+5	−5	+10
1, 2	−5	5	−10
2, 1	−15	−5	−10
2, 2	+15	5	+10

TABLE 9.11 Table of Interaction Residuals and Row and Column Effects

	Boys	Girls	Row Effect
Method 1	+10	−10	0
Method 2	−10	+10	0
Column Effect	−5	+5	

Rosnow and Rosenthal (1995) emphasize the point that an analysis of the residuals does not invalidate the absolute values of the row and column effects. These other tables help reveal how much each effect contributes to the group means. It is thus quite possible to have both a significant interaction and a significant main effect. Moreover, with this breakdown of summary data into component parts, it becomes clear how a group mean is an additive function of a grand mean plus the row effect plus the column effect plus the interaction residual. Conversely, each individual subject score can be predicted more or less accurately from a knowledge of group membership. The score is an additive function of the grand mean plus the row effect plus the column effect plus the interaction residual plus error, where error refers to the deviation between the score and the mean of group to which it belongs. This breakdown helps to explicate the logic behind the analysis of variance. The reader is further referred to Rosnow and Rosenthal (1995) for a fuller discussion of that relationship, including more guidance on how to create a table of predicted values by imposing row effects, column effects, and interaction effects on a table of mean cell values.

In summary, it is recommended that researchers subject data to multiple analyses for a fuller understanding of relationships. Those analyses should include overall main effects and, if claiming an interaction, an interpretation of the residuals.

HOW DO I SELECT THE BEST METHOD FOR ANALYZING THE PRETEST-POSTTEST DESIGN?

The analysis of change using a pretest-posttest control group design is one of the most deceptively simple but potentially complicated tasks facing the researcher. There are several statistical options. One can analyze simple change scores using a one-way ANOVA (or *t* test). One can assume that randomization takes care of pretest differences and conduct a between group ANOVA on the posttest scores. One can use a two-way repeated measures ANOVA where the within groups factor is represented by the trials. Or one might use an analysis of covariance where the pretest scores serve as the covariate and the treatment groups represent the between groups factor. Are these equally sound options? Which is recommended?

Among the first authors to tackle this issue was Frederic Lord (1967), who touted the use of raw **gain scores** as a method of change. The model underlying the use of gain scores, also called **change scores**, insists only that the pretest and posttest use similar scales of measurement, so that identical scores on the two scales have the same meaning. A major problem with using simple difference scores has been the fact that these scores usually are correlated with pretest scores. This is because the meaning of change differs at different points on a scale. For example, we would not expect a 95% free throw shooter in basketball to improve as much as a 40% free throw shooter. This distinction is lost at the level of comparing groups on raw difference scores. The point was developed in considerable mathematical detail by Cronbach and Furby (1970) in an important article. They concluded that one must rely on "residual gains" or "basefree measures of change" that have groups statistically controlled for their pretest differences. Cronbach and Furby's argument that raw gain scores are rarely useful set the stage for a lengthy series of rebuttals and proposals that has never fully abated. A recent chapter in the debate was articulated by Rogosa (1988), who disagrees with Cronbach and Furby about the unreliability of difference scores and claims that they are actually quite reliable under a wide range of moderate pretest-posttest correlations.

Residualizing a score removes from the posttest the portion that could have been linearly predicted from the pretest. Think of this as the difference between the observed and predicted value of the posttest, using a bivariate regression analysis in which the pretest score is the independent variable. Many researchers interpret the residualized score as a "corrected" measure of gain. That may be misleading, however, because the portion being discarded includes some legitimate change in the subjects between measures. Rather, the residualized score helps identify individuals who changed more (or less) than expected from their pretest scores (Cronbach & Furby, 1970). A more sophisticated residualized score was developed by Tucker, Damarin, and Messick (1966) as a "base-free measure of change." They split the difference score into two components: the part dependent on the true pretest score and the part totally independent of it. Cronbach and Furby have remarked that the Tucker et al. formulas are intended for correlational work and cannot be used to provide estimates of individual "basefree" scores because they use estimates that do not correlate with the pretest. They are intended to identify individuals whose gains are larger (or smaller) than predicted from the initial testing. As an example, imagine you are interested in identifying individuals who are especially resistant to the effects of medication (i.e., underachieve) or who are excellent candidates for leadership positions (i.e., overachieve). For these kinds of problems, the use of regression formulas is much more suitable than the use of raw change scores. This is in large part due to the "regression to the mean" problem, about which we will have more to say shortly. Thus, one can use residual gain scores or one of the formulas proposed by Cronbach and Furby (1970) for estimating an individual's true residual gain for this purpose.

? What Is Analysis of Covariance?

Treatment designs can be divided into those in which cases are randomly assigned to conditions (the true experiment) and those in which the treatment groups have not been formed at random. In both cases, according to Cronbach and Furby (1970), the use of simple difference scores can and should be avoided. In the former case, an **analysis of covariance** (with the pretest score as covariate) will take into account within group differences. ANCOVA was developed by Fisher in 1948 as a way of reducing error variance in randomized experiments. Recently, it has been employed more frequently to offer statistical control in quasi-experimental studies (where the assignment of subjects to groups may not be random). Use of the procedure has been well reviewed and explicated by Porter and Raudenbush (1987), who maintain that the analysis of covariance has been overused for this purpose and underused for the purpose of increasing the precision of statistical estimates.

To help the reader understand the logic of analysis of covariance, we present an example using a pretest-posttest control group design, with four group means, as follows.

	Pretest	Posttest
Experimental	$\overline{X}_e$	$\overline{Y}_e$
Control	$\overline{X}_c$	$\overline{Y}_c$

We represent the pretest means with X values and the posttest means with Y values. In this example, we conceptualize the pretest means as covariates. We wish to "adjust" the posttest means (i.e., the Y values) for differences in pretest

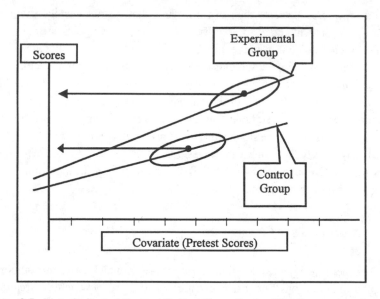

Figure 9.5. Plot of a Pretest-Posttest Control Group Design With Regression of
Dependent Variable Scores on Pretest Scores (Covariate)

means between the experimental and control groups using ANCOVA. We
present a graphic representation of this design in Figure 9.5.

In this figure, the single covariate ($\overline{X}$, the pretest score) is scaled across the
horizontal axis. The values of the dependent variable are represented on the
vertical axis, with the location of the group means indicated by the arrows
pointing toward the Y axis. The vertical distance between these arrows represents
the effect of the treatment on the experimental group. Within this figure, we have
drawn ovals to represent the spread of the cases (i.e., the scatterplot) of the
bivariate distributions of the covariate (pretest scores) and the dependent variable
(posttest scores) within each treatment group. In addition, we have drawn lines
to represent the regression of Y on X for experimental and control groups. Each
regression line passes through the mean of the respective group for which it was
calculated and the mean of the covariate for the cases within that group. This
point, shown on the graph as a dark circle, where the horizontal line representing
the group mean and the regression line meet, is called a **group centroid** and is
represented mathematically as ($\overline{X}_e$, $\overline{Y}_e$) or ($\overline{X}_c$, $\overline{Y}_c$). Note that the regression lines
are not parallel, indicating that the slopes for the regression of the dependent
variable (posttest score) on the covariate (pretest score) are not identical. To meet
the assumption of **homogeneity of regression** in analysis of covariance, how-

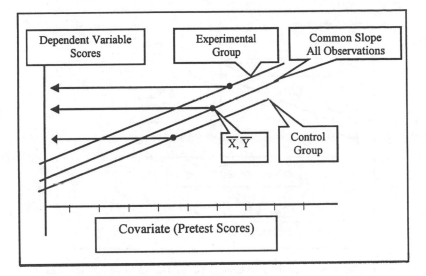

Figure 9.6. Plot of a Pretest-Posttest Control Group Design (Homogeneity of Regression Assumed)

ever, we must be able to assume that these slopes are equal. (To do so, we would fail to reject the null hypothesis that these slopes are equal.) For the purposes of this example, we will assume that these slopes are equal and redraw the experiment with equal regression slopes. We illustrate this in Figure 9.6.

Note that in Figure 9.6, we have added the line representing the regression through the group centroids of all observations, $\overline{X}$, $\overline{Y}$, and have eliminated the ovals representing the spread of cases around the group centroids.

The adjustment process is graphically illustrated, in Figure 9.7, as movement up or down each common regression line until the vertical line, representing the overall mean of the covariate, is reached. In this example, we can see that the adjustment for the experimental group is "down," whereas the adjustment for the control group is "up." The adjusted means are represented by dashed horizontal lines added to the figure. Because the mean of the control group on the covariate was relatively lower than the mean of the experimental group was higher, the required adjustment is larger for the control group. This is reflected on the graph by the larger distance between the adjusted and unadjusted group means for the control group.

Basically, an ANCOVA treatment effect is identical to an ANOVA treatment effect that has been adjusted by subtracting the product of the covariate, X, and

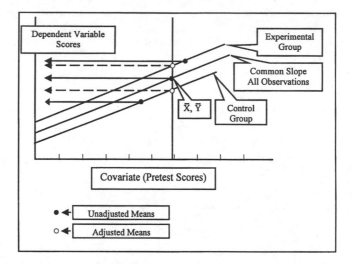

Figure 9.7. Plot of a Pretest-Posttest Control Group Design Showing Adjusted Group Means

the regression slope of Y on X, $B_{y.x}$. The ANCOVA yields a more accurate estimate of treatment effects by correcting posttest scores using the regression between pretest and posttest. By reducing the variance of the dependent variable after the treatment effect is accounted for, ANCOVA can improve statistical power and precision. In the case of randomized experiments, the interpretation is just the same as for ANOVA. When $B_{y.x} = 1$, ANCOVA is the same as a gain score analysis. Because this is rarely the case, however, the covariance analysis generally is preferable. The true controversy comes when the data are not part of randomized experiments, which is often the case in survey studies with variables that are observed in cross-sectional designs.

Porter and Raudenbush (1987) are quite dubious about the wisdom of employing the analysis of covariance with nonrandomized experiments. They much prefer avoiding nonrandomized studies in general and recommend that researchers consider path analysis and structural equation models to estimate causal relations from such data. Others are more supportive of the advantages of using ANCOVA in such circumstances, especially as the correlation between pretest and posttest drifts away from the value of 1.0. One must, however, exercise caution. Because the groups were not randomly configured, one can never be sure that they will continue to differ on preexisting variables other than the covariate. As Lord (1967) has stated, "There simply is no logical or statistical procedure that can be counted on to make proper allowances for uncontrolled preexisting differences between groups" (p. 305).

At least two sources of change may confuse the interpretation of designs utilizing a pretest-posttest control group framework. The **regression effect** refers to the well-known but poorly understood idea that a retest of cases on the extreme of a normal distribution will tend to lead to scores that are now closer to the middle. Thus, heavy people will get lighter and light people will get heavier on retesting. Regression to the mean is a bias attributed to the lack of perfect reliability of measures of variables. In practice, this means that pretest status is often the best predictor of change scores, in that it discriminates effectively between people who change positively and those who change negatively. Unfortunately, this relationship can obscure the presence of other variables (e.g., treatments or covariates) that have interesting relationships with change resulting from treatment effects. Ideally, we would like to equalize the proportion of gainers and losers at each level of the pretest, which, in turn, would be equal to the proportion of gainers and losers at every other level of the pretest. Statistically, this means calculating regression lines from the regression of the pretest (Y) on the posttest (X) and using deviations from the regression line to identify changes. This is akin to holding initial scores constant statistically in the search for correlates of change. Lord (1967) remarks that some people take this to mean that deviation from the regression line is the "real" measure of change, rather than difference scores, a point that he asserts does not make much sense. Individual raw scores are a valid indicator of change, whereas residuals can be rather confusing. It is just that researchers usually are interested in the relationship between variables when initial scores are held constant. They are interested in comparing data actually obtained with data that would have been obtained under the null hypothesis of no treatment effect.

The second primary source of confusion in change studies is errors of measurement (Lord, 1967). Scores that include errors of measurement are called **fallible**, whereas scores that do not contain errors of measurement are called **true**. When errors of measurement are unbiased (the measuring instrument works okay), they average out to 0, so that the variables approach their true values. Any measured value is only an approximation of the true value, according to classical test score theory. Thus, fallible pretest and posttest scores yield fallible difference scores. Lord (1967) offers a regression model that estimates true change from observed (fallible) values. The values obtained from regression equations always have less variability than the values being estimated; they

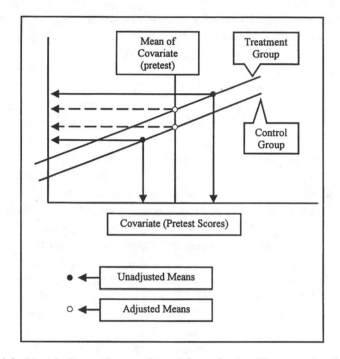

Figure 9.8. Plot of a Pretest-Posttest Control Group Design (Homogeneity of Regression Assumed)

represent the best fit of a regression line. The estimated score for a particular individual will depend both on the individual pretest score and on the group to which the individual is assigned.

To deal with regression effects, the groups must either be controlled experimentally for pretreatment differences or be controlled by using an analysis of covariance. Note that an analysis of change scores and an analysis of covariance may yield different conclusions (Lord, 1967). Figure 9.8 shows scatterplots for two groups, treatment and control, each with a pretest (X) and a posttest (Y).

The average (mean) observed scores for both the pretest (covariate) and posttest (dependent variable) are represented by arrows on the X and Y axes, respectively. The mean change for treatment group $(\overline{Y}_t - \overline{X}_t)$ and control group $(\overline{Y}_c - \overline{X}_c)$ are identical, suggesting no main effect from treatment. An analysis of covariance uses adjusted posttest scores, represented as deviations from the regression line of Y on X for each group, assuming equal slopes, as shown in the figure. This analysis shows that the means differ, unlike the gain score analysis, which is biased due to pre-group differences. This bias may not be large or significant unless the sample sizes per group are small or the groups have not been randomly composed. Without random assignment, or the ability to argue that the groups are effectively random, the researcher is hard pressed to determine the appropriate adjustment for pretest differences between groups (Lord, 1967).

O'Connor (1972) notes that analysis of covariance and multiple regression analysis suffer from the same regression to the mean phenomenon as do matching procedures. The estimated treatment effect is the sum of the actual treatment effect and the regression artifact. In a randomized experiment, matching works to reduce error as long as the matching variable does not correlate with the treatment. Because this may be difficult to determine, why not use analysis of covariance, multiple regression, or a randomized block design, which provide more information? Multiple regression also controls for initial differences. With nonrandomized, existent groups, matching is trickier, because matching on one variable often leads to undermatching on another variable. For instance, if you match on intelligence scores, you are likely to have group differences on academic achievement. Another recommendation is to use the **alternate ranks** procedure for assigning subjects to conditions in order to increase power (Maxwell, Delaney, & Dill, 1984). This procedure consists of placing subjects in groups by ranking them on pretest scores and then systematically assigning them to the various conditions so as to equalize the rankings among the groups.

? Recommendations

Of the methods used for analyzing change, the simplest is simply to ignore the pretest data and analyze the posttest data among groups. This approach is an offshoot of an experimental design known as the **posttest only** design (Campbell & Stanley, 1966) when it is implemented by not even collecting pretest data.[1] The rationale for the design is that randomization should control for between group differences, whereas the inclusion of a pretest can reduce external validity by confounding change resulting from an intervention with change resulting from pretest sensitization or the interaction of the pretest with the treatment. In practice, however, the researcher often is interested in knowing the amount of change in addition to making group comparisons and may not have enough subjects to be confident in the success of random assignment. There is simply no way of checking the success of the randomization procedure, and the presence of inadvertent bias in assignment to conditions can compromise the conclusions

TABLE 9.12 Methods of Analysis for a Pretest-Posttest Design

Method	Advantages	Disadvantages
Gain scores	Easy to compute	Do not control for pretest differences; may be unreliable
Adjusted (residualized) gain scores	Controls for pretest differences; can identify high- and low-change individuals	May discard some legitimate data
Analysis of covariance (ANCOVA)	Takes group differences into account; controls for regression effects; provides statistical control	Problematic with nonrandomized studies
Repeated measures analysis	Considers pretest scores	Does not control or adjust for pretest; interaction term is true measure of change and may be overlooked

of the study. In terms of data analysis, it is rarely a good idea to get rid of data, which is exactly what is done by ignoring pretest data that are available. Moreover, the method has the least power of the available alternatives.

The **repeated measures** approach is preferred to the posttest only method. Huck and McLean (1975) have pointed out how results are often misinterpreted by the careless researcher using this approach, however. In a repeated measures design, the main effect from a *t* test or ANOVA is not really the analysis of interest. It is the interaction effect that shows the differential impact of the two treatments across time, given the determining influence of the pretest scores. This can be confusing. The reader is referred to Chapter 9 for additional guidelines for correctly interpreting interactions.

The use of gain scores or difference scores is mathematically equivalent to the treatment × trials interaction in the repeated measures approach. That makes it more straightforward. The preceding presentation of the issues should indicate, however, that relying on raw gain scores to measure change is fraught with controversy, with proponents lining up on both sides of the issue.

It can be shown that an analysis of difference scores or a repeated measures analysis is the same as an analysis of covariance when the pretest and posttest are correlated $r = 1.0$. Because this is rarely the case, the analysis of covariance will control for pretest differences and generally provides a more powerful analysis. This is especially the case when the correlation between pretest and posttest is relatively low. Our reading of the literature indicates that most researchers prefer the analysis of covariance approach; however, the reader is referred to an article by Rogosa (1988), which provides examples of how incorrect conclusions can be generated by this method in certain atypical situations. The difference score approach is the recommended alternative when it can be easily interpreted and power already is high. As the reader can appreciate, the questions that surround the appropriate analysis of a pretest-posttest design are complex, and the issues are not resolved. In Table 9.12, we briefly summarize some of the strengths and weaknesses of different approaches.

? How Do I Assess Substantive Significance?

It has been noted that evaluating the effect of a treatment intervention by testing grouped data for statistical significance has at least two limitations (Jacobson, Follette, & Revenstorf, 1984). One is that the effects are based on average change scores for all the subjects and there is no clear way to determine the impact of the intervention on individuals. Second, as we have mentioned in a previous context, the presence of statistical significance does not necessarily imply the presence of substantive significance. Jacobson and his colleagues (1984) have addressed these issues within the context of psychotherapy outcome research, but their perceptions apply equally well to other assessments of change in experimental research. They recommend using dual criteria of change: One criterion is based on traditional significance testing of grouped data and asks whether the change is statistically reliable; the other addresses each individual subject and asks whether the scores at posttest reflect a level of functioning that is consistent with normative functioning with respect to the particular clinical problem being assessed as a dependent variable. The researcher would supplement summaries of between group comparisons with statements of proportions of "improved" clients, thus helping to explain the presence or absence of treatment effects.

Jacobson and colleagues (1984) suggest operationalizing the variable of clinical significance in one of three ways: (a) Does the level of functioning after the intervention exceed the range of the *dysfunctional* population, where the range is specified as including two standard deviations above the mean for the population (a conservative estimate of functionality)? (b) Does the level of functioning after the intervention fall within the range of the *functional* or normal population, where the range begins at two standard deviations below the mean for the normal population? or (c) Is the subject more likely to fit in the functional than the dysfunctional population based on scores on the dependent variable after the intervention? Each subject in the sample would be evaluated with regard to one of these criteria to determine significant clinical change, and these results would be expressed as a proportion of subjects who changed significantly as a result of the treatment.

The recommended analysis begins with a standard comparison of change between the experimental group and the control group(s). Then the question is asked: "For how many clients in this sample was there improvement of a sufficient magnitude to rule out chance as a plausible competing explanation?" (p. 344). A determination would be made that the change is statistically reliable. Jacobson and colleagues (1984) suggest a number of ways of making this determination but seem to prefer a **reliable change index** (RC), which is derived by dividing the pre-post difference score by the standard error of measurement:

$$RC = (\overline{X}_2 - \overline{X}_1)/SE$$

Where:

RC = reliable change score

$\overline{X}_2 - \overline{X}_1$ = posttest mean minus pretest mean

SE = standard error of mean difference.

Based on $p = .05$, an RC larger than 1.96 would be unlikely to occur in the absence of real change. One disadvantage of this measure of reliable change is that it is influenced by the reliability of the measure and can be large for small changes based on very sensitive measures. Once an individual's change is regarded to be statistically reliable, the issue of clinical significance is evaluated. It is likely that only a subset of subjects who show statistically reliable improvement will also show clinically significant change.[2]

WHAT ARE PLANNED AND POST HOC COMPARISONS?

When one conducts a one-way analysis of variance, a significant finding, implied by the size of the F ratio, indicates only that at least two means are different. If the independent variable contains only two groups, there is no problem, but what if the independent variable contains four groups? A significant F ratio does *not* tell the researcher which pair or pairs of means created by the four groups are significantly different. Because there are six pairs of means that can be formed with four groups, it is clear that much valuable information is missing. For this reason, the F ratio is sometimes called the **omnibus** or overall F ratio, meaning that it only tells us that at least one pair of means is significantly different. This situation is similar, but more complex, with factorial designs: A significant F ratio for a factor tells us only that at least one pair of means composing that factor is significant; it does not tell us which pair. It also does not tell us whether or not any one cell in the design is significantly different from any other cell. Clearly, then, we need a method for conducting the needed comparisons. Statisticians have developed two types. A **planned** comparison, also called an **a priori** comparison, refers to a specific comparison that is specified by the researcher prior to the collection of data. Another way of stating this is that planned comparisons are comparisons between the means of groups that are specified before the data are collected. This is in contrast to **post hoc** comparisons, which are tests of statistical difference between groups conducted after an overall difference has been found. The researcher is faced with determining the appropriate context for the use of either planned or post hoc comparisons. The answer to this question is very relevant to the forthcoming issue regarding error rates and multiple comparisons.

? What Are Planned and A Priori Contrasts?

As stated above, a standard ANOVA is referred to as an unfocused or **omnibus** test. The results of the test merely indicate whether something other than chance accounts for the observed differences between groups. There is no way of telling the source of any statistically significant differences without further analysis of a *post hoc* nature. The results of a *t* test can specify precisely the source of any observed group differences because only two groups and one dependent variable are involved in the analysis. Thus, a *t* test responds to a specific, focused question. In the same way, a researcher can ask specific, focused questions about the difference between any two means, and these differences can be explored with an ANOVA design. Such planned comparisons, also known as **contrasts** because the results are contrasted with specific predictions made by the investigator, have advantages over unfocused, omnibus tests.

Strictly speaking, a contrast or comparison among means is a linear combination of means that have known weights (coefficients) that sum to zero. When there are only two (nonzero) coefficients, the contrast is known as a pairwise comparison. When there are more than two (nonzero) coefficients, it is a non-pairwise comparison. In any set of p means, it is possible to generate

$p(1 - p)/2$ pairwise comparisons. For example, a three-group one-way analysis of variance produces the following three pairwise comparisons [$3(3 - 1)/2 = 3$]: $\overline{X}_1 - \overline{X}_2$, $\overline{X}_1 - \overline{X}_3$, and $\overline{X}_2 - \overline{X}_3$. In addition, we could develop comparisons involving more than two means, as follows:

$$(\overline{X}_1 + \overline{X}_2)/\overline{X}_3$$
$$(\overline{X}_1 + \overline{X}_3)/\overline{X}_2$$
$$(\overline{X}_2 + \overline{X}_3)/\overline{X}_1.$$

Let us take an example in which the researcher is investigating the impact of three psychiatric ward climates (authoritarian, benevolent, and laissez-faire) on the recovery rates of patients. A more informed researcher might anticipate and be interested in some very specific comparisons between groups and adopt the following hypotheses:

Hypothesis 1: The authoritarian ward will lead to quicker recovery than either the benevolent or the laissez-faire ward.

Hypothesis 2: The benevolent ward will lead to quicker recovery than the laissez-faire ward.

These predicted differences are weighted so that they are consistent with the theory. These weights are called **contrast weights** and the procedure is known as a **contrast analysis** (Rosenthal & Rosnow, 1991). The weights, which reflect the hypothesized differences, are represented by the Greek symbol lambda (λ). They can assume any values for a given contrast as long as their sum adds to zero. In the preceding example, the following weights would most easily accomplish this goal:

Hypothesis 1 weights: +2 (authoritarian), −1 (benevolent), and −1 (laissez-faire)

Hypothesis 2 weights: 0 (authoritarian), +1 (benevolent), and −1 (laissez-faire).

An alternative but equivalent set of weights is

Hypothesis 1: (−1, 1/2, 1/2)

Hypothesis 2: (0, −2, +2).

The second set of weights is **redundant** because it is a linear combination of the first set of weights. When contrasts are mutually nonredundant, they are called **orthogonal** contrasts. Sample weights are available for generating many

TABLE 9.13 Contrast Weights for Pairwise Comparisons Among Four Groups

Comparison	Mean 1	Mean 2	Mean 3	Mean 4	Contrast
1	1	−1	0	0	$\overline{X}_1 - \overline{X}_2$
2	1	0	−1	0	$\overline{X}_1 - \overline{X}_3$
3	1	0	0	−1	$\overline{X}_1 - \overline{X}_4$
4	0	1	−1	0	$\overline{X}_2 - \overline{X}_3$
5	0	1	0	−1	$\overline{X}_2 - \overline{X}_4$
6	0	0	1	−1	$\overline{X}_3 - \overline{X}_4$

different kinds of contrasts, including those that reflect non-linear as well as linear relationships among the variables and those that reflect combinations between means. For example, Table 9.13 represents a set of contrast weights for pairwise comparisons for four groups.

One straightforward advantage is that focused questions are more satisfying conceptually because they follow a theoretical trail derived from the researcher's hypotheses and hunches about the phenomenon being studied. Significant results from planned contrasts "make sense" because they are consistent with the thinking that went into the design of the study. Significant results from more generic, unfocused analyses are neither as precise nor as informative, and some amount of post hoc reasoning must be conducted to explain them.

Secondarily, the virtue of using focused tests is an increase in power. In exchange for hypothesizing the source of any significant findings in an ANOVA prior to analyzing the data, the researcher inherits a more powerful design. This is because planned contrasts are based on only one degree of freedom, so that the sum of squares is identical to the mean square, yielding a higher numerator for the ensuing F test. The power of *post hoc* tests, on the other hand, is low because they have to examine all possible relationships among the variables. We will have more to say about error terms for focused and omnibus tests in the following section, which deals with multiple comparisons.

? What Are Post Hoc and A Posteriori Tests?

The opposite of using *a priori* contrasts is to rely on *post hoc* tests, that is, looking for differences between groups after finding an overall difference with some omnibus test such as an ANOVA. Whereas planned comparisons are conducted in lieu of an overall F test, post hoc tests are conducted in addition to an overall F test that is statistically significant. These tests are used to pinpoint what happened in the data to account for the statistically significant finding. Because such comparisons are not specified prior to data collection by the researcher, they may be harder to justify theoretically and, hence, are regarded as more *exploratory* or, pejoratively, as "fishing expeditions." Common post hoc tests include Fisher's least significant difference test (LSD), Duncan's multiple range test, the Newman-Keuls test, Tukey's honest significant difference test (HSD), and Scheffé's general multiple comparison procedure. These pairwise multiple comparison procedures are arranged from the most liberal to the most conservative in terms of control of Type I error. They can all be computed using standard statistical software programs such as SAS® and SPSS.® All of them carry less power than a planned comparison. Although they are relatively robust with regard to small differences in group variances, alternative methods (e.g., Hartley's F-max and Levene's test) are available for data that contain large differences in group variances (Snedecor & Cochran, 1987).

A detailed examination of each post hoc test is beyond the scope of this book. The methods all seek to keep the overall significance level of the study within stated limits (e.g., .05) while allowing you to ask more questions than an analysis of variance can provide (Godfrey, 1985). Scheffé's test allows a full examination of the data prior to choosing one or more weighted combinations of means to test. This method is the least likely multiple comparison technique to identify significant differences because you are permitted to look at many more combinations than the other methods permit. Tukey's test assesses the difference among group means by referring to the difference between the smallest group mean and the largest group mean (the range) as a measure of their dispersion. This method yields narrower confidence limits than Scheffé's test. Tukey's test is especially popular for exploring all pairwise comparisons with a large number of groups. The Newman-Keuls test and the Duncan test use a different strategy. They produce clusters of group means that might hypothetically be drawn from the same population. In a five-group study, for instance, groups 1, 2, and 5 might fall into one cluster and groups 1, 3, and 4 might fall into a second cluster based on the closeness of their means. The Newman-Keuls test is more conservative than Duncan's multiple range test. Table 9.14 presents a summary of a priori and post hoc multiple comparison procedures. Each test is accompanied by the error rate most frequently associated with it, the form of comparisons it usually addresses, and the type of test.

All these multiple comparison tests have their proponents and their purposes. The more conservative methods yield broader confidence intervals, are less likely to announce a difference between means when none exists, and are more

TABLE 9.14 A Comparison of Multiple Comparison Procedures

Test	Statistical Base (test)	Comparison	Error Rate	Comments
A priori tests				
Individual t tests	t	Pairwise	Per comparison	Inflated alpha level
Linear contrast	F	Any contrast	Per comparison	When orthogonal, greatest power for simple and complex contrasts
Bonferroni	t	Any contrast	Familywise	Can make all possible simple and complex planned comparisons
Post hoc tests				
Fisher's least significant difference (LSD)	t	Pairwise	Familywise	Assumes significant F; controls familywise error indirectly
Newman-Keuls	q	Pairwise	Familywise	Good use for all-around test
Duncan	q	Any contrast	Contrast	Not recommended because of unsteady alpha levels
Tukey	q	Pairwise	Familywise	Robust with unequal Ns; good power for pairwise contrasts
Scheffé	F	Any contrast	Familywise	Very little power for testing multiple hypotheses
Dunnett's	t	With control	Familywise	Control vs. all others; can't make other comparisons

likely to overlook a significant comparison (Godfrey, 1985). Conversely, the less conservative methods have greater power: They yield narrower confidence intervals and are more likely to detect small differences but may also claim differences that do not hold up as well. For a more complete comparison among the methods, we refer you to a review article by Jaccard, Becker, and Wood (1984). Those authors provide recommendations for choosing the best post hoc test for various between groups, within groups, and mixed designs.

It should be clear that planned comparisons are preferable to post hoc comparisons whenever possible. The exception is in cases of more open-ended, exploratory research, and you will pay some price in terms of reduced power and higher error rates. One can also combine planned and post hoc comparisons. The typical approach is to test the significance of a few orthogonal planned comparisons of particular interest, then conduct an overall F test on the residual treatment sum of squares, followed by post hoc tests if the F is significant. If you are using an ANOVA design, Rosenthal and Rosnow (1991) advise you to plan t tests that are of interest to you prior to data collection and conduct those t tests whether or not the overall F test is significant. In fact, one need not compute an overall F test at all; the only rationale for doing so is to yield a more stable estimate of the variance for the denominator of the t tests. The authors go on to state that if the t tests reveal some interesting, unanticipated results, one can then compute the overall F. If the F is significant, it offers protection against the possibility that the t tests were significant by chance, because some comparisons must be significant if the overall F is significant (Snedecor & Cochran, 1980).

HOW DO I CONTROL FOR FAMILYWISE ERROR?

The issue of error rates and multiple comparisons is one of the more vexing considerations in data analysis. Previously, we have discussed the use of alpha levels in hypothesis testing to control for Type I errors. Assume a typical study where we have set the alpha level at .05, meaning that we are willing to accept a 5% chance of making a Type I error in a comparison between two groups. Many studies, however, are much more ambitious and involve **multiple comparisons** or a large number of statistical tests. Multiple comparisons, sometimes called **post hoc comparisons**, refer to the possible comparisons that can be identified in a factorial or ANOVA design. An example not involving analysis of variance is computing t tests on every pair of a large set of independent variables in an attempt to find a statistically significant difference or examining a correlation matrix to identify the statistically significant correlations (presumably at the same .05 level). Note that with only 10 variables, we are speaking of 45 possible pairwise comparisons.

These examples are noteworthy, not because they are recommended (they are not) but because they are distressingly common. The problem, of course, is that these multiple comparisons are treated as if they were independent of one another, which is a problematical assumption, and they can be expected to result in a few significant differences purely on the basis of chance. Take the example of a researcher who is assessing allergic skin reactions in 30 subjects by presenting each of them, in randomized single-subject sessions, with 10 different detergents. The resultant data consist of 10 dependent variables and 30 subjects, leading to 300 separate comparisons. The probability of making at least one Type I error at the .05 significance level is, of course, .05 ($p = 1 - .95 = .05$). This probability rises swiftly with multiple comparisons. For 10 comparisons, it becomes $(1 - .95^{10}) = 0.401263$. In this study, it becomes highly predictable that several comparisons will be significant by chance: $(1 - .95^{300}) = .999997$. Thus, it is a virtual certainty that when a large number of tests is conducted, at least some Type I errors will be made. The researcher needs to consider the impact of multiple comparisons on Type I error so that statistical conclusions may be interpreted correctly. The researcher also needs to be aware of available procedures for minimizing these error rates.

? What Are Familywise Error Rates?

Each kind of test, omnibus or focused, has a probability of error associated with it. The probability of Type I error for a focused test is called the **comparison-wise** error rate, whereas the probability of Type I error for a group of tests, such as provided by ANOVA or MANOVA, is called a **familywise** or **experimentwise** error rate. Another way of stating this is that a comparison-wise error rate refers to the degree of risk assigned to a single dependent variable; an experimentwise error rate refers to the degree of risk that accrues in the analysis of more than one dependent variable per experiment (Haase & Ellis, 1987). In the first situation, each comparison is used as the unit for computing the error rate; in the second situation, the entire experiment becomes the unit for determining the error rate. The more comparisons made, the greater the probability of obtaining chance errors and the higher the experimentwise error rate. The formula for this computation with independent tests is

$$\alpha_{EW} = 1 - (1 - \alpha_{PC})^{P}$$

Where:

α_{EW} = experimentwise alpha

α_{PC} = error rate for a single comparison (comparison-wise alpha)

P = the number of comparisons (number of dependent variables).

Thus, with only five tests, the experimentwise error rate would be $1 - (1 - .05)^5 = .23$.

Philosophically, there are those who maintain that familywise error inflation is not worth considering (Darlington, 1990; Saville, 1990). They argue that the natural unit is the individual comparison, not the group or family of comparisons that make up an experiment or study, and that it makes no sense to have the scope of the study penalized because the investigator is testing several hypotheses. As Saville (1990) puts it, one might as well select all the statements in a statistician's lifetime as the appropriate family! They especially do not like to see researchers punished for conducting ambitious multifactorial studies. They claim that there is no difference between a study that consists of 20 comparisons and 20 separate studies with one comparison each. There is some logic to this argument, yet the majority of contemporary researchers seem to side with the other position, namely, protecting the experimentwise alpha rate, which means keeping the error rate at the alpha level no matter how many comparisons are conducted. This also seems to be the position of most professional research journals, although recent reviews have found that a distressingly large number of published articles contain inflated Type I error rates that have not been adequately addressed (Dar et al., 1994).

? Recommendations

Methods to protect the familywise alpha rate depend on the design of the study. Where comparisons are truly independent, it makes sense to employ a more conservative alpha level by dividing .05 (or whatever level of significance is being used) by the number of comparisons being made. If, for instance, there are 10 independent tests, the alpha level would be reduced to .005. Differences that exceed this level of significance could be used confidently to reject the null hypothesis. The name of this *a priori* multiple comparison method is the **Bonferroni correction**. It has also been called the Dunn Multiple Comparison Test. The Bonferroni approach changes the criterion for individual comparisons so that the probability of finding at least one significant comparison for the entire experiment is maintained at the original *p* value (e.g., .05). The Bonferroni correction reduces the power of the study but should be a satisfactory strategy when there are not many differences to compare. (Note, however, that some critics of null hypothesis significance testing might claim that the Bonferroni correction adjusts for non-existent alpha error.) When comparisons are not truly independent but are part of the same overall analysis, other multiple comparison methods are available (cf. Einot & Gabriel, 1975; Jaccard et al., 1984; Kirk, 1996). Recall that Table 9.14 offers a summary of several commonly employed multiple comparison techniques. The table summarizes the type of comparison being made and its statistical base. We also offer some brief comments about each test. Note that there is no totally fair method to compare these techniques because each was designed for a different purpose.

As we have stated previously, the more tests of significance are computed on data for which the null hypothesis is true, the more significant results will be found and the more Type I errors will result. The value of doing planned comparisons instead of *post hoc* ones is that the overall alpha level in planned comparisons needs to be divided only by the number of comparisons that are anticipated. If there is one comparison to consider, planned in advance, the *t* test works well, as does a linear contrast for a more complex comparison. With several planned comparisons to consider, the revised alpha has to be formed by dividing the overall alpha by all the possible pairwise comparisons to eliminate the chance of bias. The Bonferroni method can be used very flexibly because one need not set the alpha rate the same for all comparisons. What is important is leaving the familywise alpha rate intact (e.g., .05). This alpha can be divided among the anticipated comparisons any way we choose. If a particular *t* test is of great importance to the study, one can impose more power on it, perhaps by invoking a .02 level of significance and dividing the remaining .03 among the other comparisons.

If you choose to assign importance ratings to planned comparisons in this way, it is critical to do so in advance of analyzing the data so that the ratings are not made as a function of the results that are obtained. Even in the case of multiple comparisons, a finding that has a low probability of occurrence (e.g., $p < .001$) will stand out. Dividing this value (.001) into the experimentwise alpha rate (.05) can tell us how many comparisons would have to be made before one would question the significant result as being due to chance ($N = 50$) (Rosenthal & Rosnow, 1991). A general rule of thumb suggests that one should test only those contrasts that are relevant to the theoretical and/or applied purpose of the study. But this is probably not in line with what researchers actually do and would result in the absence of exploratory data analyses that might uncover many potential research leads. Our suggestion is to make clear in advance those tests that are based on a priori reasoning or specific hypotheses. These can be used as the "set" of findings within which the experimentwise alpha level will be distributed. Then acknowledge those additional analyses based on an exploratory approach or on questions generated by the more restricted set of hypothesis-based findings, as subject to Type I error. Including both types of findings in the same "pool" for alpha adjustment potentially would make the alpha level for any one comparison so low that we would drastically raise the level of Type II error and severely jeopardize the likelihood of achieving any statistical significance.

WHEN SHOULD I USE MANOVA?

Research studies frequently address the relationships among independent variables and multiple dependent measures within the same design. For example, imagine a research project examining the influence of grade level, student gender, and teaching method on several dimensions of student performance. In such a study, it might be reasonable to consider the performance measures as a "set" of conceptually related dependent variables and include them in a MANOVA, or multivariate analysis of variance. MANOVA, however, is a complex and sophisticated analytical method that should be used with caution, and it probably is overused. In the following paragraphs, we provide an overview of this method and some practical advice on its use.

The availability of powerful applied statistical packages makes this kind of data crunching relatively easy. The commonly accepted analysis strategy begins with a multivariate analysis of variance (MANOVA), a follow-up of significant effects with univariate ANOVAs, and then a follow-up of significant ANOVAs with appropriate multiple comparisons or contrast analyses for simple effects. The most frequently used test statistics to evaluate MANOVA hypotheses are Wilks's lambda, Pillai's trace, Hotelling's trace, and Roy's method. These measures serve as criteria for evaluating differences across the dimensions of the dependent variables. The first three assess all sources of difference among the groups regardless of whether they differ on a linear combination of the dependent variables. They can be approximated by an F statistic, which makes them relatively easy to compute and understand. Wilks's lambda may be the most popular of the measures, although Pillai's trace works better with small sample sizes, unequal cell sizes, and violations of homogeneity of covariances (Hair et al., 1995). The fourth measure, Roy's method (also called Roy's greatest characteristic root), tests the differences among the groups on only the first canonical root (discriminant function) among the dependent variables. When the dependent variables are strongly intercorrelated, so that they represent a single underlying dimension, Roy's method is particularly powerful. When several dimensions need to be considered, however, or when there are significant violations of the underlying assumptions of MANOVA, it probably should be avoided.

The analysis strategy described above has been recommended for many years (Cramer & Bock, 1966). Its rationale is based on the belief that the omnibus MANOVA provides familywise error protection for the subsequent ANOVAs, that the ANOVAs provide error protection for the multiple comparisons, and so on. In other words, the thinking has been that by using an alpha of .05 for the MANOVA and obtaining a significant effect, the researcher is now free to search the data for the location of that effect, while being comfortably "protected" from the Type I error problems associated with initiating multiple univariate tests. Recently, this logic has come under attack.

? What Are MANOVA and ANOVA?

What first needs to be noted is that MANOVA and ANOVA address very different research issues. An ANOVA explores the relationship between one or more independent variables and a single dependent variable. Thus, multiple ANOVAs examine group differences on more than one dependent variable without considering the relationship of the dependent variables to one another. In fact, univariate analyses assume that the intercorrelations among dependent variables are zero. When the dependent variables are intercorrelated, the differences between groups will be magnified with a series of univariate comparisons. For instance, if two dependent variables measure the same (significant) thing, ANOVAs will identify two statistically significant effects.

In contrast, a MANOVA considers the dependent variables simultaneously and forms groups based on *composites* of them. These composites are called **canonical variates** and are created using equations called **linear discriminant functions** (Huberty & Morris, 1989). The weighting of variables in these functions is achieved to maximize group differences on the composites. The meaning of discriminant functions is determined by the loadings (weights) of the dependent variables on them (Kaplan & Litrownik, 1977). Dependent variables that do not contribute to the discrimination between groups are assigned very little weight in the composite. The functions are adjusted on the basis of the correlations among the dependent variables. If two variables correlate highly, the effect will be split between them.

Let us take the example of comparing two groups on the variables of vigilance, listening skills, emotionality, and voice quality. A series of ANOVAs will tell you if the groups differ on each of these measures without considering the relationships among them. A MANOVA will tell you if the groups differ from each other on linear composites of these measures. It may be assumed from the MANOVA that the individual measures are related to one another conceptually. In this case, perhaps they reflect a "social sensitivity" construct and the researcher, based on theoretical considerations, may be interested in knowing if the two groups differ from one another on social sensitivity. Thus, applying a MANOVA or a series of ANOVAs to the same data set depends on the aims of the researcher.

The preceding description implies that MANOVA is especially appropriate when the researcher is interested in addressing a multivariate hypothesis, a hypothesis based on a system of variables. A variable system has been defined by Huberty and Morris (1989) as "a collection of conceptually interrelated variables that, at least potentially, determines one or more meaningful underlying variates or constructs" (p. 304). It is entirely appropriate to follow up a significant MANOVA with subsequent analyses to explore the full complexity of the multivariate data. Two kinds of appropriate research questions have been dubbed by Huberty and Morris (1989) as the "variable selection problem," identifying subsets of outcome variables that account for group differences, and the "variable ordering problem," establishing the relative contribution of the outcome variables to group differences or the treatment effects. Tatsuoka (1971) recommends discriminant analysis as a method for determining the dimensionality of the underlying data and the relationships among the dependent variables. This use of discriminant analysis in the context of MANOVA is sometimes overlooked because of the more common application of discriminant analysis for classifying individuals within specific populations (Bray & Maxwell, 1982). The respective contributions of the variables to group discrimination can be further determined with additional multivariate contrasts. In such cases, the Bonferroni procedure for controlling the familywise alpha rate to the nominal alpha level of the specific contrasts can be employed.

The preceding use of MANOVA suggests that the dependent variables of the study are intercorrelated. There is considerable disagreement in the literature whether higher correlations among the variables increase or decrease the power of the technique to find group differences. A recent article by Cole, Maxwell, Arvey, and Salas (1994) discusses this controversy. In a nutshell, the authors conclude that if one chooses dependent variables with large effect sizes (e.g., .5), power increases as the correlations among variables decrease and become negative. If one chooses dependent variables with very different effect sizes, power increases as correlations become more positive (if positive) or more negative (if negative). Of course, power generally can be increased if the researcher wants to test a specific *a priori* linear combination of dependent variables with, for example, an ANOVA. The nuances of the relationships between power and the intercorrelations among the dependent variables in a MANOVA can be found in the Cole et al. (1994) article.

This description of appropriate uses of MANOVA does not fully answer the concerns of researchers who have relied on MANOVA to control for experimentwise error rates for subsequent analyses. What is the status of this approach? There is no disagreement about the appropriateness of using MANOVA to answer intrinsically multivariate questions, such as evaluating how a set of dependent measures differs as a whole across groups. Take the common example of evaluating an intervention on different, related outcome dimensions, such as behavioral measures of change, self-report measures of change, and physiological measures of change. Many current textbooks (e.g., Hand & Taylor, 1987) recommend the use of MANOVA to exercise control over experimentwise error rate for separate univariate questions. Although it is common for researchers to begin their analysis with a MANOVA, and then follow up with ANOVAs, this strategy is regarded as problematical (Huberty & Morris, 1989). For one thing, MANOVA tables can be almost impenetrable to read and the results uninterpretable. For another, a significant MANOVA does not necessarily imply the presence of a significant ANOVA. (See Tatsuoka, 1971, for a simple bivariate example.) A repeated measures design is one example of a situation that is appropriate for analysis by either ANOVA or MANOVA but can yield contradictory results from the two approaches. In fact, an intervention that has a minimal effect on two uncorrelated dependent variables may have a larger, overlooked effect on the composite of the two variables (Kaplan & Litrownik, 1977). Note, however, that a significant MANOVA merely means that some particular combination of dependent variables distinguishes the groups, which generally is not very relevant to the researcher who is interested in individual variables. Although the intention of controlling Type I error through a MANOVA is laudatory, the alpha level for each subsequent ANOVA is less than the alpha for the MANOVA only when the MANOVA null hypothesis is true, which generally is not the time researchers adopt this strategy. In general, following up a significant MANOVA with univariate tests and specific contrasts results in a loss of power and an increase in Type I error.

? Recommendations

Huberty and Morris (1989) are very clear that multiple ANOVAs, as opposed to the omnibus MANOVA, are appropriate for answering questions such as, "With respect to which outcome variables do the groups differ?" or "Which outcome variable is affected by the treatment variable?" Most situations that mandate the use of several ANOVAs boil down to the following two categories:

1. The dependent variables in the study are conceptually independent and the researcher has no interest in looking at a linear composite of them or trying to assess an underlying construct that incorporates all of them, or

2. Exploratory research in studying the impact of an intervention on new outcome variables.

Note that any empirical interrelationships among the outcome variables are completely ignored when conducting multiple ANOVAs. This is not an issue when the outcome variables are conceptually independent, but to the extent that the variables are correlated, the analyses will yield some redundant information.

The general recommendation operating here pertains to the previously discussed advantages in power and Type I error reduction in using planned comparisons rather than *post hoc* comparisons. It is generally better to control for the error rate at the outset of the analysis. One suggested method is to apportion the alpha at the level of **families** of hypotheses as discussed previously (note the term "familywise error rate") (Dar et al., 1994). In the context of MANOVA, families of variables are variables that are conceptually linked or refer to the same construct. Sometimes they are different measures of the same construct, such as speech measures, behavioral ratings, and galvanic skin response measures of public speaking anxiety. Ultimately, the decision of what constitutes a family is a matter of judgment, and this judgment may be somewhat capricious and arbitrary (Dar et al., 1994). Once families of variables or measures are defined, the experimentwise alpha (e.g., .05) could be adopted for each family and partitioned, according to the Bonferroni correction method, within each family. This procedure is a bit of a compromise between trying to minimize Type I error and preserving the overall power of the study.

For example, let us assume that a multivariate study contains seven dependent variables that can be lumped into three relatively independent categories: three measures of economic growth, two measures of physical health, and two measures of family stability. This study could conceivably reflect three general research questions: How does the adoption of a parliamentary government affect (a) economic growth, (b) community participation, and (c) family stability? In spite of the possibility that all seven measures are intercorrelated, an alpha of .05 could be assigned to each family to maintain a desired level of power for the study. Within each family, the alpha would be appropriately partitioned (.025, .017, .017) to reduce the Type I error rate. This is consistent with Dar et al.'s (1994) recommendation to assign the adopted Type I error rate to each variable family and perform univariate analyses within each family by dividing the error rate by the number of variables in the family.

Another solution for the use of univariate follow-up tests to significant MANOVAs has been prescribed by Bird and Hadzi-Pavlovic (1983). They note that most researchers treat these two stages as unrelated. Indeed, the problems of "protected" univariate tests can be avoided if the same test statistic is used throughout the analysis. The key is to use a simultaneous test procedure (STP) derived from the multivariate test statistic used for the initial overall test with all follow-up tests. Bird and Hadzi-Pavlovic (1983) have compared several MANOVA test statistics as the basis for an STP to see which perform well at both stages of the analysis. An STP gives a way of performing tests on components of the general null hypothesis in such a way that the order of testing is irrelevant. Otherwise, we have the following scenario: In the general null hypothesis, Type I error is limited to alpha by any two-stage procedure that requires a significant result at Stage 1 before follow-up testing is allowed. If the general null hypothesis is false, you cannot achieve a Type I error with the MANOVA, but you can at Stage 2 if any testable implication of the general null hypothesis is true, whether or not the general hypothesis is true. So, we must abandon univariate tests if the purpose of the overall test is to control the maximum familywise error rate.

If the STP is based on the MANOVA statistic used for the overall test, one cannot retain the null hypothesis and also reject particular hypotheses implied by it. The familywise error rate is controlled directly by the STP, which reaches its highest level when the general null hypothesis is true. When protected univariate tests produce an inflated Type I error rate, the corresponding error rate from any MANOVA STP is smaller than alpha. In short, with such a two-stage analysis, the overall test is theoretically unnecessary because it can be omitted without affecting subsequent contrasts, but it is useful in practice because if it is nonsignificant, there is no point in going further.

So what MANOVA test statistic should be chosen? If power and robustness of the overall tests are the only criteria, then Bird and Hadzi-Pavlovic (1983) recommend the Pillai trace statistic V (Pillai, 1955). Roy's largest root statistic, R (Roy, 1953), however, always provides more power than any other MANOVA STP in tests of contrast null hypotheses in the second stage of testing and performs acceptably well in the first stage of a coherent analysis. This decision is based on considerations of power and controlling the familywise error rate below the nominal Type I error rate. It should be acceptable unless the number of contrasts to be tested seems to justify a Bonferroni-adjusted univariate STP with a smaller alpha.

Many studies are justified in incorporating more than one dependent variable. Two measures of a single trait often make sense. Moreover, adding independent variables to a study implies the need to increase the total N, whereas adding dependent variables is not as costly. In such cases, MANOVA may be used appropriately. Wording of the hypotheses to reflect the combination of the dependent variables is very important. It is recommended that you set all dependent variables in the same direction to facilitate interpretation of the findings. Because MANOVA assumes some intercorrelation among the dependent variables, you may want to determine the extent of intercorrelation with a test such as the Bartlett Test of Sphericity. The test also assumes that the dependent variables are normally distributed and have equal variances.

In conclusion, we recommend that less experienced researchers avoid the use of MANOVA and stick with ANOVA unless the following criteria are met.

1. The researcher has developed hypotheses suggesting interrelationships between several dependent variables and is interested in their combined influence.
2. The researcher has at his or her disposal consultants who will be able to discuss both the interpretation and the presentation of the findings.

We recommend that researchers should not conduct MANOVA simply because they have multiple dependent variables or because they believe that a MANOVA will "protect" the analysis from "experimentwise" error. The danger of misusing this technique is probably greater, and there are other, equally effective, methods for accomplishing this goal.

NOTES

1. Campbell and Stanley's (1966) book is a classic and includes an excellent discussion of the internal and external validity considerations associated with various pretest-posttest research designs. It is not, however, a statistics book.

2. A very different approach to analyzing pretest-posttest data across time is the use of time-series designs. Methodological issues regarding the application of these designs have been reviewed in a special issue of the *Journal of Consulting and Clinical Psychology* edited by Gottman and Rushe (1993). Most of the designs refer to analyzing a single variable recorded periodically over time, such as rainfall in Palm Springs, or the effects of a treatment intervention on a single behavior therapy client over the course of treatment. Using a *t* test or analysis of variance with single-subject data by comparing baseline observations with intervention phase observations vio-

lates the independent observations assumptions of the techniques (Crosbie, 1993). The t or F becomes artificially inflated because the observations correlate more highly with previous observations than with the mean. An interrupted time-series analysis controls for this autocorrelation factor (Gottman, 1981). Basically, the technique compares the slope and intercept of each phase of the treatment using both a t test and an F. Crosbie (1993) claims to have developed an interrupted time-series analysis (ITSA-CORR) that improves on Gottman's by controlling more effectively for Type I error with positive autocorrelations with short series. The reader is referred to the cited references for further elaboration of time-series designs.

CHAPTER 10

Questions About
Multiple Regression Analysis

What is . . . ? What are . . . ?

How do I . . . ? When should or shouldn't I . . . ?

Construct an interaction term?

Determine the appropriate number of subjects?

Interpret an interaction effect?

Interpret beta weights?

Interpret bivariate correlation coefficients?

Interpret hierarchical regression?

Interpret multiple correlations?

Interpret part and partial correlations?

Interpret regression coefficients?

Interpret stepwise regression?

Select the appropriate multiple regression design?

Use adjusted R^2 instead of R^2?

Use hierarchical methods?

Multiple regression analysis?

The relationship between predictor variables, sample size and power?

The number of predictor variables I can use?

Stepwise multiple regression?

Hierarchical multiple regression?

The nuts and bolts of MRA?

Multiple correlation?

Multiple coefficient of determination?

R^2?

A Venn diagram?

Shrinkage?

Multicollinearity?

Adjusted R^2?

Unstandardized regression coefficients?

Standardized regression coefficients?

Beta weights?

Partial correlations?

Semipartial correlations?

Level: Intermediate/Advanced

Focus: Conceptual/Instructional

Questions About Multiple Regression Analysis

The technique of multiple regression has its intellectual roots in the pioneering work of Sir Francis Galton, a prolific 19th-century genius (Cowles, 1989). Inspired in part by his cousin Charles Darwin's earthshaking publication of *Origin of Species*, Galton pursued a developing commitment to the hereditary basis for talent, power, and intellect. His research began with a study of the size and weight of sweet pea seeds over two generations. This led to early evidence for the phenomenon of **regression**, called "reversion" by Galton, indicating that the mean weight of offspring seedlings was not as extreme as the weight of parental seedlings. Later on, he ambitiously recorded physical characteristics of thousands of human volunteers and found a similar "regression" toward mediocrity in human hereditary stature (Cowles, 1989). The mathematics used to describe these relationships between intergenerational physical characteristics such as height and weight were the regression lines and slopes now associated with regression analysis. It was a short hop from Galton's final contribution to heredity and regression, the book *Natural Inheritance* (1889), to subsequent developments in correlational methods by Walter Weldon, Karl Pearson, and others. Ironically, the term "regression" is still used, based on its role in history, whereas in contemporary statistics, regression is usually more accurately noted by the term "prediction."

WHAT ARE THE NUTS AND BOLTS OF MULTIPLE REGRESSION ANALYSIS?

Multiple regression is used for analyzing data when the researcher is interested in exploring the relationship between multiple continuously distributed independent variables and a single dependent variable.[1] Historically, the independent variables are usually referred to as predictor variables and the dependent variable is referred to as the criterion variable (Share, 1984). As we shall see, however, prediction is only one of the possible uses of multiple regression. The method takes its name from the fact that it examines the regression of the dependent variable (Y score) on the multiple independent variables (X scores). A linear combination of independent variables is achieved that maximizes the multiple correlation (R) between these independent variables and the dependent variable. This linear combination is described by weightings (regression weights or regression coefficients) of the independent variables. There are two equations that represent the basic multiple regression model. The first describes the regression model itself, and the second, called the **regression prediction equation**, describes the prediction of the dependent variable or criterion variable (Y) based on the independent variables composing the regression analysis. These equations, for three independent variables, are shown below.

$$\text{Regression Model: } Y = a + b_1X_1 + b_2X_2 + b_3X_3 + e$$

$$\text{Regression Prediction Equation: } Y' = a + b_1X_1 + b_2X_2 + b_3X_3$$

In these equations, the a represents the intercept, which is the value of Y when all the X values are zero. The b values represent unstandardized regression coefficients that "weight" the independent variables (the X values) during the regression analysis. The two equations differ in the fact that the regression model includes an error or residual term, e, and the prediction equation does not. The Y value in the prediction equation includes a "prime" (Y'), indicating that this is a predicted value of Y. (Sometimes the regression line is referred to as the "Y prime line.") Regression analysis accomplishes this task by determining the b

values that weight the independent variables in such a way as to maximize the linear correlation between predicted and observed Y values. Said another way, regression analysis maximizes the correlation between the observed and predicted values. Said still another way, regression analysis minimizes the sum of the squared deviations between the observed and predicted Y values. This notion is the basis for the term **least squares**, represented by the equation

$$\Sigma(Y - Y')^2 = \Sigma e^2 = \text{minimum.}$$

The **multiple correlation coefficient** is an index of how well the scores that are predicted from the linear combination correspond to the actual scores of the dependent variable. Said another way, the multiple correlation coefficient and its square, the **multiple coefficient of determination** (R and R^2, respectively) assess the "fit" of the regression equation to the data that it describes. When the fit is perfect, both R and R^2 are 1.00 and all points will fall on the regression line (i.e., $\Sigma(Y - Y')^2 = \Sigma e^2 = 0$).

The key to the interpretation of regression analysis is understanding the interpretation of the various "coefficients" that are produced. We have already mentioned the multiple correlation coefficient and its square, the multiple coefficient of determination (R and R^2, respectively). In addition to these two, which assess the overall fit of the regression equation, there are an additional two, the unstandardized and standardized regression coefficients (b and beta, respectively). These describe the role of the individual independent variables that compose the multiple regression equation. Finally, there are a number of other correlations, including the original bivariate correlations, between the independent variables and dependent variable, and among the independent variables themselves, that should be included within the basic set of coefficients utilized in the interpretation of multiple regression analysis. We discuss these, and other issues of multiple regression analysis, in the following sections.

One of the advantages of multiple regression analysis is that it is able to examine relationships in which the independent variables are correlated with each other and with the dependent variables. This might be the case, for instance, if we were pursuing the relationship among three predictor variables—patient assertiveness, level of illness, and family income—and waiting time for receiving a medical appointment. In this chapter, we utilize Venn diagrams to express visually the relationship between variables. In these diagrams, each circle represents the total variability of the variable it designates. The overlap between circles represents shared variance and is thus a visual representation of the

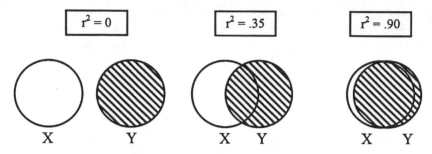

Figure 10.1. Venn Diagrams Illustrating Common Variance in Bivariate Correlation

r^2 value. Figure 10.1 illustrates the use of Venn diagrams in interpreting the meaning of the value of r^2.

The Venn diagram in Figure 10.2 illustrates one possible set of correlations between the individual predictor variables, the correlations among the predictor variables, and the resulting R^2. Each circle represents the total variability of the variable represented, and overlapping circles represent overlapping, or shared, variability.

As shown in Figure 10.2, family income and patient assertiveness are correlated, as shown by the overlap between the circles. Both of these variables are also correlated with the dependent variable. Some of the overlap with the dependent variable, however, is redundant. The manner in which this overlap is attributed to the variables is dependent on the type of multiple regression being conducted; however, this variability is **never** counted twice. Note also that the strongest predictor of waiting time, level of illness, is not correlated with either of the other two independent variables and thus does not overlap with these two variables on the Venn diagram. We discuss the different manner in which different approaches to regression analysis address overlapping or shared variability in a later section of this chapter.

HOW DO I DETERMINE THE APPROPRIATE NUMBER OF SUBJECTS AND PREDICTOR VARIABLES?

It is quite easy to become enamored with the capability of multiple regression analysis to deal with multiple independent variables, but what are the limits? Can one include 5 independent variables, or 10, or 15? The answer to questions concerning the number of independent variables that is "reasonable" depends

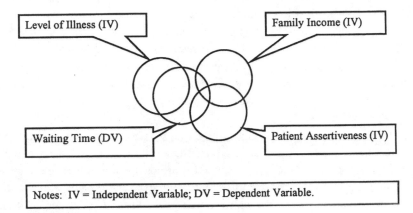

Notes: IV = Independent Variable; DV = Dependent Variable.

Figure 10.2. Venn Diagram Illustrating the Assignment of Variability in Multiple Regression

on a number of factors, most important among them the number of subjects, but one should also consider the potential effect size, the desired power, and the alpha level (level of significance) to be used. This question becomes particularly relevant when a large instrument has been administered to a small sample.

Theoretically, you may have as many independent variables as you want, but the loss in power can be substantial when the ratio of number of subjects to number of variables becomes smaller than 15. One way to think about this issue is in terms of the concept of **shrinkage**, the difference between the R^2 based on data from one sample and the R^2 derived from the same regression equation applied to a second sample. For example, if we calculated the value of R^2 on one sample, then applied the derived regression weights to calculate the value of R^2 on another sample from the same population, the second R^2 would almost certainly be smaller. This "shrinkage" occurs because the calculation of regression weights to obtain a maximum R^2 always capitalizes, to a certain extent, on chance (unless the bivariate correlations are error free, which is virtually never the case). When R^2 is .50, validity shrinkage begins to become small when the ratio n/k is 15 and k is the number of predictors (about 12% of the sample R^2). With more subjects, very little shrinkage is likely to occur.

There are a number of rules of thumb for approximating the appropriate number of subjects. One is that the number of subjects necessary to test the multiple correlation is less than the number necessary to test the individual predictor variables. The simplest rule is to apply the formula $N \geq 50 + 8k$, where k is the number of independent variables, for testing the multiple correlation, and $N \geq 104 + k$ for testing individual predictors. Thus, with five independent variables the recommended sample sizes would be $50 + 8(5) = 90$ and $104 + 5 = 109$. With 10 independent variables, the recommended sample sizes would be 130 and 114. When testing both the multiple correlation and the individual predictors, which is almost always the case, calculate both numbers and select the larger one. The above rules assume an alpha level of .05, a power of .8, and a medium effect size.

What sorts of situations might suggest an increase in these recommended sample sizes? First, if the effect size is small, the power level will drop quickly without an increase in the sample size. This means that the researcher runs the risk of failing to detect an effect (i.e., committing a Type II error) if he or she utilizes a rule of thumb for larger effect sizes. Second, evidence of poor reliability in the variables indicates the presence of measurement error that will attenuate correlations. Thus, a larger sample size will be necessary. Finally, if the dependent variable is not normally distributed, you will need to either transform the variable or increase the number of subjects.

The above suggestions relate to standard regression in which all independent variables are entered into the regression in one step or group. One might expect that if stepwise regression is being used, the number of subjects required would decrease, because the stepping process eliminates the redundant or poorly associated variables. Unfortunately, this is exactly the opposite of what one should do. Because stepwise procedures tend to capitalize on chance variation in the data, they tend to produce solutions that are not replicable, meaning that they are not repeatable and do not generalize beyond the data at hand. Tabachnick and Fidell (1996) suggest a ratio of cases to independent variable of 40 to 1 with stepwise procedures.

To provide the reader some rough guidelines, we have produced Table 10.1, which presents recommended sample sizes for two power levels (.6 and .8), at an alpha level of .05, for three effect sizes (small, medium, and large), with three, five, and seven independent variables.

WHAT ARE STEPWISE AND HIERARCHICAL MULTIPLE REGRESSION PROCEDURES, AND WHEN SHOULD I USE EACH?

There are three major multiple regression procedures, and researchers often are confused about the differences among them and the appropriate context for their use. These different procedures can be distinguished in terms of both how variability is partitioned among the variables and who controls the process of

TABLE 10.1 Regression Sample Sizes for Combinations of Independent Variables, Power, and Effect Size

Number of Independent Variables	Power	Effect Size		
		Small	Medium	Large
3	.6	362	52	25
3	.8	549	77	36
5	.6	436	64	31
5	.8	648	92	43
7	.6	495	73	36
7	.8	726	104	49

entering variables into the regression equation. Being clear about the goal of the research helps to determine the appropriate approach. Researchers typically have one of three different goals in mind (Darlington, 1968; Rawlings, 1988; Wapold & Freund, 1987). Each of these goals has implications regarding the way in which the independent variables are included or entered into the regression equations.

The first goal is to describe the relationships between the independent and dependent variables. Here the **simultaneous entry** of all the variables into the analysis is appropriate. This is generally considered the standard multiple regression technique. The researcher can then evaluate the statistical significance of R^2 without worrying about eliminating variables or considering causal explanations. This is a very modest research goal, but it does allow the researcher to assess the multiple correlation between a dependent variable and a group of independent variables and to assess the impact of each independent variable, controlling for the others.

The second possible goal is prediction. The independent variables are treated as predictors, and the researcher seeks to determine the best subset of them to predict the criterion at a high level of accuracy. The goal is to maximize R^2 (or some other statistical criterion that may be maximized or minimized) while minimizing the number of predictors. For example, both automobile and life insurance companies make regular use of regression to predict such criterion variables as the likelihood of an auto accident and longevity. Thus, we know that unmarried, young males are more likely to be involved in auto accidents than married, middle-aged men. When we are attempting to predict something like longevity, there are many potential variables to consider, including gender, race, eating habits, socioeconomic status, health status of parents and siblings, drinking and smoking habits, and distance driven to work. What is the optimal combination of predictors? The appropriate procedure here is stepwise or "setwise" regression. In one version, **all subsets regression**, all possible subsets of predictor variables are tried to answer this question. The primary alternative is **stepwise regression**, where the order of entry of the independent variables into the regression equation is determined by a selection algorithm. In these procedures, the researcher is, in one sense, removed from the analysis because the computer utilizes purely statistical criteria to enter and remove variables from an analysis. Thus, as we will see below, these procedures have often been subject to criticism. A "forward selection" procedure starts by entering the variable that has the highest correlation with the criterion. The next variable entered is the one that contributes the largest increase to R^2, beyond the R^2 contributed by the first variable. Another way of putting this is that the second selected variable has the highest first-order partial correlation with the criterion. (We define partial and other types of correlation below.) The forward selection procedure continues in this manner until some preselected criterion is reached, such as the absence of a statistically significant increase in R^2.

Another stepwise procedure is called "backward elimination." It is the logical opposite of the forward selection procedure. All the predictor variables are entered into the equation at the outset. The variable that contributes the least increase to the total R^2 is removed first, and so on until some criterion is reached.

The most commonly employed stepwise procedure capitalizes on the advantages of both adding and subtracting predictor variables. The procedure automatically adds a variable or subtracts a variable depending on which step would most significantly enhance the model (i.e., increase the R^2). Short of the all subsets regression, this approach will yield the model that best fits the sample data. In general, stepwise procedures are controversial and not highly recommended. The model development is atheoretical, and the results tend to "overfit" the data because they take advantage of chance relationships in the sample. Moreover, the significance tests based on the stepwise procedures (F) are misleading because they ignore the larger number of predictor variables that have been introduced to the model but not retained (Wampold & Freund, 1987). Cohen and Cohen (1983) have concluded that stepwise multiple regression, a very popular procedure, should be conducted only when (a) the research goal is primarily to predict a criterion, (b) the sample size is large (a ratio of about 40:1 for subjects to predictor variables), and (c) the results are cross-validated on a new sample.

The third goal of multiple regression is testing a theoretical model. Under these circumstances, the researcher chooses the order of entry of variables into the equation based on some theoretical rationale. This is called **hierarchical regression**. Generally, the variables that are entered first are those that are regarded as (a) being particularly important or previously determined to relate to the dependent variable, or (b) confounding the analysis. In the first case, the researcher might be interested in the proportion of variance accounted for by a variable above and beyond the contribution of the variables previously entered. In the second case, the variables entered are viewed as "covariates" that the researcher wishes to remove from the analysis prior to assessing the contribution of the variables that are of theoretical importance. Thus, the answer to the question, "How do I control for the confounding influence of extraneous factors in multiple regression?" is to remove the variability due to those factors by entering them first into a hierarchical procedure.

The anticipated nature of the relationships among the variables should help determine their order of entry into the equation. Let us say that a researcher thinks that the influence of alcohol on sociability is mediated by level of confidence. That is, confidence is a *mediator* variable that is affected by alcohol intake and that, in turn, increases sociability. The model looks like this:

$$\text{Alcohol} \rightarrow \text{Confidence} \rightarrow \text{Sociability.}$$

Another researcher insists that alcohol directly increases sociability but admits that confidence can also influence sociability. This model takes the following form:

The order of entering the variables into the regression analysis becomes crucial in testing these models. To test the first hypothesis, the researcher first enters the data on confidence level into the equation. Sociability is the criterion variable. Amount of alcohol intake would be entered subsequently. If the first hypothesis were true, one would expect that alcohol would not contribute a significant amount of variance in sociability when confidence was in the model; in the case of the second hypothesis, alcohol should add significantly to the prediction of the criterion, and one variable would not take precedence over the other. Any overlap in the explained variability due to sociability and alcohol

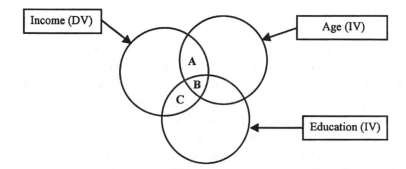

Figure 10.3. Assigning Variability in Standard and Hierarchical Multiple Regression

consumption would contribute to the total variance explained but would not be attributed uniquely to either variable.

The following is a more detailed example comparing standard multiple regression analysis with hierarchical multiple regression analysis. Consider the relationship between age, level of education, and income. In assessing the effect of education on income, we would want to control for the effects of age, because age and income are correlated, as are age and education. As people age, they move further into their careers and presumably make more money. (In fact, age and income may interact to predict even higher levels of income for those who are well educated. We discuss interaction effects in regression analysis in a later section.) A standard multiple regression analysis will assess the effect of age on income, controlling for education, and the effect of education on income, controlling for age. These effects would be revealed in the standardized and unstandardized regression coefficients, and the R^2 would reflect the total variability explained by the two variables. If, however, we wanted to remove all the variability in income associated with age, prior to assessing the effect of education, we would run the analysis in a hierarchical fashion, entering age first, then income. The Venn diagrams in Figure 10.3 represent the differences in the assignment of variability in the two different multiple regression analyses.

TABLE 10.2 Correlation Matrix of Income, Education, and Age

Correlations

		Income Group	Ed Group	Age Group
Pearson Correlation	Income Group	1.000	.402**	.481**
	Ed Group	.402**	1.000	.260
	Age Group	.481**	.260	1.000
Sig. (2-tailed)	Income Group		.004	.000
	Ed Group	.004		.071
	Age Group	.000	.071	
N	Income Group	49	49	49
	Ed Group	49	49	49
	Age Group	49	49	49

**Correlation is significant at the 0.01 level (two-tailed).

In the standard analysis, the unique variability due to age (A) would be assigned to age and the unique variability due to education (C) would be assigned to education. The overlapping variability (B) would be included in the total R^2 value, but would not be assigned to either age or education. In the hierarchical procedure, the overlapping variability would be assigned to age in the first step (A + B), and the variability unique to education (C) would be assigned to education. We illustrate these principles below with a numerical example.

? A Detailed Example:
The Effect of Age and Education on Income

The correlation matrix in Table 10.2 presents correlations among age, education, and income. Don't worry about the exact values of the raw scores; the data have been partly grouped into age, education, and income categories but still represent continuous data.

The correlation matrix shows that both age and education are correlated with income, and that age and education are positively correlated (.260). We present

TABLE 10.3 Model Summary for Regression of Income on Age and Education

Model Summary[a,b]

Model	Variables Entered	Variables Removed	R	R Square	Adjusted R Square	Std. Error of the Estimate
1	Age Group, Ed Group[c,d]		.560	.314	.284	2.18

a. Dependent Variable: Income Group
b. Method: Enter
c. Independent Variables: (Constant), Age Group, Ed Group
d. All requested variables entered.

three different analyses of the data, one standard and two hierarchical. Beginning with the standard multiple regression model summary in Table 10.3, the R^2 value (.314) shows that age and education, together, explain 31.4% of the variance in income.

Next, to examine the significance of the regression model, we examine the ANOVA summary table (Table 10.4). The summary table shows that the results are statistically significant and the total regression (explained) sum of squares is 99.827. Note that we can reproduce the value of R^2 by dividing 317.837 into 99.827. When the significance level is given as .000, it means that the values are less than .0005. We typically recommend that the value be reported as < .001, because a probability of .000 makes no sense.

Following the assessment of the overall equation, we examine the individual regression coefficients, shown in the table of coefficients in Table 10.5. The standardized regression coefficients show that the effect of age, controlling for education (.404), is larger than the effect of education, controlling for age (.297), and both effects are statistically significant. Note that the unstandardized coefficient for age is given in scientific notation. This can be read as .091 (E-02 means that 9.1 is multiplied by 10 to the power of −2).

In the hierarchical analysis model summary shown in Table 10.6, we entered age in the first step and education in the second. Note that this is reflected in the printout as Model 1 and Model 2. The R^2 for Model 1 is .232, indicating that 23.2% of the variance in education is explained by age, the only variable in the

TABLE 10.4   ANOVA Summary for Regression of Income on Age and
Education

ANOVA[a]

Model		Sum of Squares	df	Mean Square	F	Sig.
1	Regression	99.827	2	49.914	10.532	000
	Residual	218.010	46	4.739		
	Total	317.837	48			

a. Independent Variables: (Constant), Age Group, Ed Group

equation at this step. In the second model, we add the effect of education, and
the R^2 increases to .314, the same value shown for the R^2 in the standard analysis
and indicating that we have been able to explain an additional 8.2% of the
variance (i.e., .314 − .232). What is different is that we have now attributed
almost 3/4 (73.88%) of this variability to the effect of age and only 1/4 to
education.

The ANOVA summary table and coefficient table are shown on pages 260
and 261, (Tables 10.7 and 10.8), and for Model 2 they are identical to the ones
seen previously. Changing the method of entry of the variables does not change
the explained sum of squares when the variables included in the equation are the
same. Similarly, when both variables are included in the equation (Model 2), the

TABLE 10.5   Coefficient Summary for Regression of Income on Age and
Education

Coefficients[a]

Model		Unstandardized Coefficients B	Unstandardized Coefficients Std. Error	Standardized Coefficients Beta	t	Sig.
1	(Constant)	2.505	1.166		2.149	.037
	Ed Group	.369		.297	2.351	.023
	Age Group	9.1E-02	.028	.404	3.196	.003

a. Dependent Variable: Income Group

TABLE 10.6   Model Summary for Hierarchical Regression of Income on Age and Education

Model Summary[a,b]

Model	Variables Entered	Removed	R	R Square	Adjusted R Square	Std. Error of the Estimate
1	Age Group[c,d]	.	.481	.232	.215	2.28
	Ed Group[e]	.	.560	.314	.284	2.18

a. Dependent Variable: Income Group
b. Method: Enter
c. Independent Variables: (Constant), Age Group
d. All requested variables entered.
e. Independent Variables: (Constant), Age Group, Ed Group

standardized and unstandardized regression coefficients are identical to those obtained in the standard analysis.

Finally, for the sake of illustration, we redo the analysis in hierarchical fashion, entering education first, and then age. The results are shown in Table 10.9, beginning with the model summary.

TABLE 10.7   ANOVA Summary for Hierarchical Regression of Income on Age and Education

ANOVA[a]

Model		Sum of Squares	df	Mean Square	F	Sig.
1	Regression	73.641	1	73.641	14.174	000[b]
	Residual	244.195	47	5.196		
	Total	317.837	48			
2	Regression	99.827	2	49.914	10.532	000[c]
	Residual	218.010	46	4.739		
	Total	317.837	48			

a. Dependent Variable: Income Group
b. Independent Variables: (Constant), Age Group
c. Independent Variables: (Constant), Age Group, Ed Group

TABLE 10.8 Coefficient Summary for Hierarchical Regression of Income on Age and Education

Coefficients[a]

Model		Unstandardized Coefficients		Standardized Coefficients		
		B	Std. Error	Beta	t	Sig.
1	(Constant)	4.214	.955		4.414	.000
	Age Group	.108	.029	.481	3.765	.000
2	(Constant)	2.505	1.166		2.149	.037
	Age Group	9.1E-02	.028	.404	3.196	.003
	Ed Group	.369	.157	.297	2.351	.023

a. Dependent Variable: Income Group

As we would expect, the total variance explained is the same (31.4%), but the findings now show that education, entered first, explains 16.2% of the variance in income, and age explains 15.2%. Thus, the percentage of the total variability attributed to education and income is about 50% each, rather than the 3/4 to 1/4 distribution shown earlier. Note also that, as in the other two analyses, the standardized and unstandardized regression coefficients are identical in the

TABLE 10.9 Model Summary for Hierarchical Regression of Income on Education and Age

Model Summary[a,b]

Model	Variables			R Square	Adjusted R Square	Std. Error of the Estimate
	Entered	Removed	R			
1	Ed Group[c]	.	.402	.162	.144	2.38
2	Age Group[d,e]	.	.560	.314	.284	2.18

a. Dependent Variable: Income Group
b. Method: Enter
c. Independent Variables: (Constant), Ed Group
d. All requested variables entered.
e. Independent Variables: (Constant), Ed Group, Age Group

TABLE 10.10 Coefficient Matrix for Hierarchical Regression of Income on
Education and Age

Coefficients[a]

Model		Unstandardized Coefficients		Standardized Coefficients		
		B	Std. Error	Beta	t	Sig.
1	(Constant)	4.547	1.067		4.263	.000
	Age Group	.499	.166	.402	3.012	.004
2	(Constant)	2.505	1.166		2.149	.037
	Ed Group	.369	.157	.297	2.351	.023
	Age Group	9.1E-02	.028	.404	3.196	.003

a. Dependent Variable: Income Group

two variable models (Table 10.10). Thus, what is different is not the total variance explained, or the values of the regression coefficients, but the manner in which the overlapping variability is attributed to the two variables: either as a joint contribution of both, with the combined total of 31.4%, or 3/4 due to age and 1/4 due to education as in the first hierarchical analysis, or as about equally due to income and education, as in the second hierarchical analysis. Note that if we had chosen to run a stepwise analysis, either forward or backward, the results would have been the same as the first hierarchical regression model. This is because the bivariate relationship between income and age (.481) is stronger than the bivariate relationship between income and education (.402). This means that age would have been entered first in a forward stepwise procedure, or eliminated last in a backward stepwise procedure.

WHAT ARE REGRESSION COEFFICIENTS
AND HOW DO I INTERPRET THEM?

Regression analysis produces a lot of printout. In fact, even for seasoned researchers, the sheer volume of printout can be overwhelming. For a relative novice, the choices of what to include and what to exclude, what to interpret and what to ignore, and what it all means can result in a crisis of confidence in one's ability to ever comprehend this material. We have found that one way to assist with the interpretation of regression analysis output is to conceptualize this material as belonging in three separate, but highly connected, groups based on the concept of "coefficient." A **coefficient** is a number that describes a relationship or can be manipulated in such a way that it describes a relationship. Thus, we organize our three groups of coefficients to reflect three different parts of the interpretive process that one is likely to find useful in the description of regression analysis.

The first of these groups is composed only of bivariate coefficients. These express the bivariate relationships between the independent and dependent variables to be considered for inclusion in, or utilized in, a multiple regression analysis, as well as the relationships between the independent variables themselves. Thus, we call our first group "coefficients that describe bivariate relationships," and only two coefficients are included, the bivariate r and its square (r^2), the coefficient of determination.

The second group is composed of "coefficients that describe the overall regression equation." These are measures of "fit" of the regression equation to the linear or non-linear model being tested. Again, there are only two coefficients to be considered, the multiple correlation (R) and its square, the multiple coefficient of determination (R^2).

The third group is the largest and is composed of "coefficients that describe the role of the individual predictor variables in the regression analysis." These include the standardized (beta) and unstandardized (b) regression coefficients, the partial and the semipartial correlation.

Note first that each of the coefficients we have mentioned above can be tested for statistical significance. We recommend any of the references we have cited for these formulas. Second, we have excluded from this discussion coefficients that relate to the testing of assumptions prior to engaging in a multiple regression. We have discussed these elsewhere, particularly in Chapter 5. Our concern here is with interpretation of completed analyses in which the assumptions have been met.

? How Do I Interpret Bivariate Correlation Coefficients?

There are two coefficients that are central to our consideration of multiple regression analysis: the bivariate r and its square (r^2), the coefficient of determination. (We reserve capital R and R^2 for the multiple correlation and its square.) Both r and r^2 contain the same information, though the value of r^2 will always be positive and can be interpreted as the amount of variance in the dependent variable Y that is explained by the independent variable X, or simply as the

TABLE 10.11   Interpreting Bivariate Correlation Coefficients

Correlation Coefficient (r)	Interpretation	Coefficient of Determination (r^2)
1.00	Perfect positive	1.00
.80	Strong positive	.64
.50	Moderate positive	.25
.20	Weak positive	.04
.00	No relationship	.00
−.20	Weak negative	.04
−.50	Moderate negative	.25
−.80	Strong negative	.64
−1.00	Perfect negative	1.00

amount of shared variance. Table 10.11 provides some guidelines for the interpretation of bivariate correlation coefficients.

In the context of regression analysis, we suggest that one always begin with an examination of the bivariate relationships. This is critical for a number of reasons. First, an examination of the relationship between the independent variables detects overlapping variability. If this is large, which it can be when variables measure identical or highly overlapping concepts, problems of **multicollinearity** may be present. Multicollinearity, or high correlations between independent variables, or composites of independent variables, can be problematic for regression solutions, and you may need to consider eliminating one of the variables, forming a composite, or analyzing each separately. (For a discussion of how to diagnose and adjust for multicollinearity, see Tabachnick and Fidell, 1996, p. 84.) Because of the manner in which multiple regression controls for overlapping variability, there may be different reasons why a variable may appear relatively unimportant in a multiple regression analysis. One reason may be the fact that the variable is not correlated with the dependent variable. An equally plausible reason may be that the variable is highly correlated with a number of independent variables included in the regression model. An examination of the bivariate relationships will make this clear. We illustrate two possible outcomes with Venn diagrams in Figure 10.4.

In the first outcome, the relationship between the dependent variable (DV) and independent variable 2 (X_2) would appear weak in a multiple regression analysis because the bivariate relationship is weak. In the second outcome, the

Independent variable (X_2) weakly
related to dependent, little overlap.

Independent variable (X_2) strongly
related to dependent, large overlap.

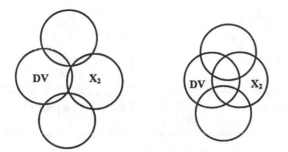

Figure 10.4. Two Scenarios Illustrating Apparently Weak Relationships Between
Dependent and Independent Variables in Multiple Regression

bivariate relationship is strong, but the multiple regression equation would most
likely show little or no effect of this variable because almost all of its shared
variability with the dependent variable is overlapping with X_1 and X_3. To
understand this phenomenon, we would need to examine the correlation matrices
among the independent variables and between the independent and dependent
variables.

? How Do I Interpret Multiple Correlations?

Just as a bivariate correlation describes the relationship between one independent and one dependent variable, a **multiple correlation** describes the
relationship between two or more independent variables and one dependent
variable. We use capital R and R^2 to describe these relationships. In a Venn
diagram, the total variance explained, represented by R^2, the **multiple coefficient
of determination**, is the total overlap between all the independent variables and
the dependent variable. When the independent variables are correlated, as they
almost always will be, variance that is shared by two or more independent
variables is not counted twice. The concept of shared variability between
independent variables that is also shared with a dependent variable is illustrated
in Figures 10.4 and 10.5. The figures show two Venn diagrams, one with
considerable overlap among the independent variables and one with very little
overlap. Thus, we can think of the relationship between the dependent variable

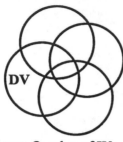

Large Overlap of IVs

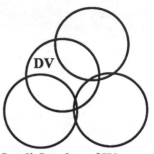
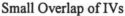
Small Overlap of IVs

Figure 10.5. Illustration of Overlapping Variance in Two Multiple Correlations

and the independent variables as being either relatively independent of the relationship of the independent variables with one another or highly confounded.

Most computer programs provide both an R^2 value and an "adjusted" R^2 value. Why is it necessary to adjust this value? The answer is that as we continue to add independent variables into a regression equation, even if all the variables are unrelated to the dependent variable, we would be accumulating chance variation, which is all in the positive direction (the values are all squared). For example, four independent variables, none of which are related to the dependent variable, might by chance correlate (bivariate r) $-.2$, $-.3$, $+.1$, and $+.4$. Even though these average to zero, when they are squared and summed, the value is .30 (assuming that the independent variables are not correlated with each other). Thus, the value of R^2 may be artificially inflated, particularly if the analysis is utilizing a stepwise variable entry routine. In addition, the smaller the sample, the greater the magnitude of these chance fluctuations. The formula for adjustment provided by Tabachnick and Fidell (1996) is

$$\overline{R}^2 = 1 - (1-R^2)\left(\frac{N-1}{N-k-1}\right)$$

where N is the sample size, k is the number of independent variables, and R^2 is the squared multiple correlation. With four independent variables and a squared multiple correlation of .3, as in our example above, the adjusted R^2 would be .188 with 30 subjects and .291 with 300 subjects. Although the value of R^2 was reduced by about 37% in the 30-subject case, some authors (cf. Cattin, 1980) suggest that with small sample sizes an even greater adjustment is necessary.

? What Are Unstandardized and Standardized Regression Coefficients?

Probably the biggest confusion regarding the interpretation of regression analysis concerns the two coefficients that are used to "weight" the individual predictor variables. These are the standardized and unstandardized regression coefficients, or *b* and beta, respectively. There is little agreement about the exact label to apply to the standardized regression coefficient.[2] The interpretation of the **unstandardized regression coefficient** is that it represents the amount of change in the dependent variable associated with a one-unit change in that independent variable, with all other independent variables held constant. The unstandardized regression coefficients are the optimal values by which to multiply predictor values to obtain estimates of criterion values. Thus, they are useful for making actual predictions.

A couple of warnings are in order. First, *b* values are also highly scale dependent, and thus unsuitable for measuring the "importance" of a predictor variable within any one equation. Said a bit more strongly, the unstandardized regression coefficients tell us very little about the individual importance of a variable in a regression equation relative to the other variables. These weights are the optimal coefficients for predictor variables to obtain estimated criterion scores. The regression coefficient, however, depends on the unit of measurement of the independent and dependent variables. The unstandardized regression coefficient for the effect of temperature on frustration would be different, for example, if temperature were being measured using a Celsius scale instead of a Fahrenheit scale. Second, the values of the unstandardized (and standardized) regression coefficients change as a function of the other variables in the equation. If, for example, we eliminated from a regression equation a variable that produced overlapping variability with both the independent variable in question and the dependent variable, the unstandardized regression coefficient would change, perhaps tremendously. Thus, unstandardized regression weights can be very deceptive measures of the importance of predictor variables.

When the various independent variables rely on different measurement scales, **standardized regression coefficients**, or **beta weights**, are used to compare the individual predictor variables within an equation. Think of these values as representing the solution in which all the variables have been converted to standard scores. This permits the comparison of independent variables measured on very different scales, for example, income, measured in thousands of dollars per year; education, measured in number of years; and motivation, measured on a 10-point scale. Rather than comparing scale scores with dollars and years, standardization allows us to compare variables with a mean of zero and a standard deviation of 1. Thus, the interpretation of beta weights is in terms of the expected change in the dependent variable, expressed in standard scores, associated with a change of one standard deviation in an independent variable, while holding the remaining independent variables constant. Although the interpretation of *b* and beta within a regression equation seems reasonably clear, additional problems of interpretation may arise when we attempt to compare these values across different samples and populations. According to Kerlinger and Pedhazur (1982):

> It is important, however, to bear in mind that the magnitude of a beta reflects not only the presumed effect of the variable with which it is associated but also the variances and the covariances of the variables included in the model, as well as the variance of the variables not included in the model and subsumed under the error term. Because some, or all, of these factors may vary from one population to another, it is possible for a given variable to have a relatively large beta in one sample and be declared important and a relatively small beta in another sample and be declared unimportant or less important. In short, *beta is sample-specific and can therefore not be used for the purpose of generalizations across settings and populations.*
>
> The unstandardized coefficient, on the other hand, remains fairly stable despite differences in the variances and the covariances of the variables in different settings or populations.
>
> Because of the relative stability of the b's in different populations, most authors prefer them over the betas as indices of the effects of the variables with which they are associated. (pp. 248-251, emphasis added)

Thus, in line with Kerlinger and Pedhazur's recommendations and that of many other authors, we recommend the use of beta weights when assessing the effects of different variables within a single regression equation or population. When the purpose is to compare populations, even with similar regression equations, our recommendation is to compare the unstandardized coefficients.

Finally, the above comments are important when considering what should be reported when presenting the results of a multiple regression analysis. We recommend, first, that the bivariate correlations of all variables, both independent and dependent, be reported, as well as their means and standard deviations. Second, we recommend that both the standardized and unstandardized regression coefficients and the standard error of the unstandardized coefficient be presented. Third, we recommend that the researcher consider presenting the partial correlation as an aid to interpreting the role of the independent

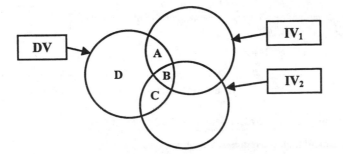

Figure 10.6. Assigning Variability in Different Regression Frameworks Using Bivariate, Partial, and Semipartial Correlations

variables in the regression equation. We discuss the partial and semipartial correlation coefficients below.

Thus far, we have discussed the meaning of six different coefficients in the interpretation of multiple regression analysis. The use of beta weights helps us understand the importance of an independent variable, relative to the other independent variables in an equation, but is *not* a measure of the unique variance explained. To get a sense of this unique variability, we turn to the partial or semipartial correlation. This can be illustrated with the use of Venn diagrams. In the three-variable model shown in Figure 10.6, both the independent variables are equally correlated with the dependent variable, and part of this shared variability overlaps with the dependent variable.

Partitioning of variability for bivariate squared correlations (r^2)

$r^2_{yx1} = (a + b)/(a + b + c + d)$
$r^2_{yx2} = (b + c)/(a + b + c + d)$

Notes and comments: The bivariate correlations are independent of the relationships between the independent variables.

Partitioning of variability for semipartial squared correlations (r^2)

Standard multiple regression analysis
$sr^2_{yx1} = (a)/(a + b + c + d)$
$sr^2_{yx2} = (c)/(a + b + c + d)$

Notes and comments: Unique variability in the independent variables is related to the total variability in the dependent variable.

Hierarchical multiple regression analysis (X_1 entered first)

$sr^2_{yx1} = (a + b)/(a + b + c + d)$

$sr^2_{yx2} = (c)/(a + b + c + d)$

Notes and comments: All shared variance of X_1 with the dependent variable is included. Only unique variability with X_2 is included.

Partitioning of variability for partial squared correlations

Standard multiple regression analysis

$pr_{yx1} = (a)/(a + d)$

$pr_{yx2} = (c)/(c + d)$

Notes and comments: Overlapping and unique variability in the independent variable and dependent variable is considered. All variability shared with other independent variables is eliminated.

Hierarchical multiple regression analysis (X_1 entered first)

$pr^2_{yx1} = (a + b)/(a + b + d)$

All shared variability with X_1 is included. Variability shared between X_2 and the dependent variable is eliminated.

$pr^2_{yx2} = (c)/(c + d)$

Notes and comments: Overlapping and unique variability between X_2 and the dependent variable is considered. All variability shared with other independent variables is eliminated.

To summarize the patterns shown in Table 10.6, the bivariate correlations reflect only the shared variance between the dependent and independent variables. The type of analysis conducted does not change the value of these correlations. Both the partial and semipartial correlations must be interpreted in light of the type of regression analysis being conducted. In **partial correlation**, we remove the overlapping variability created by the presence of other independent variables from both the independent and the dependent variables. In **semipartial correlation**, we remove the overlapping variability from the independent variable only. Thus, with semipartial correlations, the total variability is consistently the total sum of squares around the dependent variable. With partial correlation, this total can vary, as a function of both the variables being considered and the type of analysis. This consistency in the nature of the total variability for the semipartial correlation suggests that it may be a more useful measure and less prone to misinterpretation. We can think of the semipartial correlation as representing the *unique* variability contributed by that independent variable in that regression equation. In that equation, the semipartial correlation represents the amount by which the R^2 would be reduced if that variable were removed from the regression equation.

HOW DO I INTERPRET
AN INTERACTION EFFECT?

When conducting ANOVA, the issue of interaction typically is taken care of by the computer program itself and thus becomes transparent, meaning that the researcher may not explicitly need to build interaction effects into the analysis. In fact, the researcher may need to limit the number of interactions produced in complex factorial designs. In multiple regression analysis, on the other hand, the issue must be addressed directly by the researcher. Unfortunately, a number of authors (Jaccard, Turrisi, & Choi, 1990; Kerlinger & Pedhazur, 1982) have observed that interaction terms are often absent in multiple regression analysis, even when there may be very good empirical and theoretical reasons for including them. In Chapter 9, we discussed three components of the study of interaction. The first was to determine whether or not an interaction effect exists in the population. If we determine that an interaction exists, the second question concerns the strength of the effect, usually expressed in terms of some measure of effect size or variance explained. The third component was to address the exact nature of the interaction in terms of both its empirical understanding and its theoretical relevance.

These three goals are exactly the same in multiple regression, but they are approached in a slightly different manner. For one thing, we are typically dealing with continuous variables, not categorical variables as in analysis of variance. Second, the interaction term must be deliberately included in the regression equation by the researcher. Finally, the explanation of the meaning of the interaction typically is more difficult when working with regression models. For this last reason, in particular, researchers who have been interested in evaluating interaction effects have fallen prey to the trap of dichotomizing or trichotomizing the chosen variables and conducting an ANOVA. Kerlinger and Pedhazur (1982) and others have strongly recommended against this approach, and we have discussed some of the problems created by this practice in Chapter 8.

This section addresses how to include and evaluate interaction terms in a multiple regression analysis. We deal primarily with continuous variables, but interactions between categorical variables (dummy coded) and interactions between continuous and categorical variables also can be addressed in this context. A good discussion of both dummy coding methods for use in multiple regression and the examination of interaction terms with dummy coded and continuous variables is available in Kerlinger and Pedhazur (1982).

In the multiple regression model specified below for the two-variable case, we assume a basic linear additive model, meaning that the effects of each variable can simply be added together to predict the value of the dependent variable, Y.

$$Y = a + b_1X_1 + b_2X_2 + e$$

It also makes sense to ask, "Because there are two independent variables involved, isn't it possible, even reasonable, that these variables might interact?" The answer to this question is "Yes." Just because the variables may be continuous does not preclude their interaction. A set of variables in which an interaction exists, but is not included within the model, represents a case of "misspecification." (We have discussed the problems with model misspecification in Chapter 5.) Our advice is that you should include within your analytical model all the variables (including interaction terms) that are likely to affect a relationship.

The process of including an interaction term in a regression analysis is quite direct. A new variable is created by multiplying the values of the two variables that are thought to interact. With our two-variable model above, the new model, including the interaction terms, would be

$$Y = a + b_1X_1 + b_2X_2 + b_3X_1X_2 + e.$$

The first question we address relative to interaction terms is whether or not they exist. In the above analysis, this question would be answered by calculating the three-variable regression prediction equation and examining the statistical significance of the regression coefficient b_3. If there are more than two independent variables, one may include terms for additional interaction effects, and if one wishes to examine three-way or higher order interactions, additional multiplicative terms, created by multiplying three or more variables together, can be created. Thus, to examine the existence of an interaction in multiple regression, you must create a variable representing that interaction and examine its statistical significance. There is some debate regarding what to do if the interaction term is not significant. Eliminating the nonsignificant term creates a more parsimonious model, increases the degrees of freedom in the residual sum of squares, and thus increases power, even if only slightly. Nonsignificance of an interaction term, however, may be attributable to low power and, thus, a theoretically important piece of information may be lost. It is up to the researcher to consider the role of the interaction in the analysis, the power of the design, and the potential loss of information created by deleting an interaction term solely on the basis of its statistical significance.

The second question concerns the strength of the effect. We showed in Chapter 9 that we could calculate the eta^2 value for the interaction term in factorial ANOVA by dividing the total sum of squares by the sum of squares attributable to the interaction term. In multiple regression analysis, we have two options. The first is to run the analysis twice, once excluding the interaction term and once including it. We then subtract the lower value of R^2 from the higher value. The difference can be attributed to the effect of the interaction term. Even though formulas are available for addressing the significance of the difference between two R^2 values, we do not need to address the question of whether or not this difference is statistically significant in this manner. The t test of the regression coefficient, b, is equivalent to this test and typically is provided by standard multiple regression statistical packages. This excludes the need for additional hand calculations.

The second manner in which we might address the question of the effect size of the interaction term is by running a hierarchical regression. The first step would include the independent variables, and the second step would add the interaction term, or terms. Examination of the R^2 change statistic, and the significance of this change, would be equivalent to running two independent regressions and subtracting the R^2 values.

Assume that we have successfully completed all the analyses that we have suggested above. We have found a statistically significant interaction, and we know how much of the variance in the model can be attributed to this term. The task is now to explain the nature of this interaction. Unfortunately, the number of types of potential interaction is very large. We refer here to what is called a **bilinear interaction**, which is only one of the possible functional forms that an interaction between continuous variables can assume. (We will mention the others briefly in the following paragraphs.) In this form, the slope between one independent variable and the dependent variable changes as a function of the scores on another independent variable. An example might help. Imagine that we are examining sexual permissiveness as a function of age and religiosity with the following multiple regression equation:

$$Y'(\text{Sexual Permissiveness}) =$$
$$3 - 2(\text{Age}) - 4(\text{Religiosity}) - 1(\text{Age} \times \text{Religiosity}).$$

If we assume that the coefficient for the interaction term (-1) is statistically significant, the slope of age on sexual permissiveness changes as a function of the value of religiosity. The nature of this change is described by the value of the slope coefficient on the interaction term (Age $\times$ Religiosity). This coefficient indicates the number of units that the slope of age on sexual permissiveness changes, given a one-unit change in religiosity. Because this value is -1, the slope of sexual permissiveness on age decreases one unit for every one unit that religiosity increases.

How can we determine the slope of sexual permissiveness on age at various levels of religiosity? Let's assume that religiosity is measured on a 10-point scale and that we want to evaluate the slope of sexual permissiveness on age at low, medium, and high values of religiosity. We first select these values as 1, 5, and 10, respectively. Any values can be used, but the intention here is to examine the effect throughout the range of religiosity. We utilize an equation provided by Jaccard et al. (1990) for evaluating the predicted slope (b_1) of one independent variable (X_1) on the dependent variable (Y) at any value of a second independent variable (X_2):

$$b_1 \text{ at } X_2 = b_1 + b_3 X_2.$$

We utilize the religiosity values of 1, 5, and 10 as follows:

$$b_1 \text{ at } 1 = \quad -2 + (-1)(1) \ = -3$$
$$b_1 \text{ at } 5 = \quad -2 + (-1)(5) \ = -7$$
$$b_1 \text{ at } 10 = -2 + (-1)(10) \ = -12.$$

The three slope coefficients above elucidate the meaning of the slope coefficient on the interaction term. (These coefficients can also be tested for statistical significance by dividing them by their standard error. Formulas are available in Jaccard et al., 1990.) The interaction coefficient indicates that for every single unit that religiosity increases, the slope of sexual permissiveness on age decreases one unit. Thus, at low levels of religiosity the slope of sexual permissiveness on age is -3. When we increase the level of religiosity by 4 units, to 5, the slope decreases 4 units, to -7. Finally, when we increase religiosity to its highest value, from 5 to 10, the slope decreases another 5 units, from −7 to −12. This is, in essence, the nature of an interaction of the bilinear functional form.

How do we interpret the regression coefficients for the independent variables in the presence of an interaction effect? A number of authors have discussed the difficulty of interpreting these coefficients (e.g., Blalock, 1970; Smith & Sasaki, 1979; Wright, 1976). Jaccard et al. (1990) argue that these coefficients are interpretable, but they are interpreted differently from coefficients without an interaction term because they in fact estimate different concepts. In a typical model containing no interaction effects, we interpret the unstandardized regression coefficient b_1 as representing the amount of change in Y for each unit change in X_1, holding the value of the other independent variables constant or controlling for their effects. Thus, according to Jaccard et al. (1990), these coefficients "estimate general relationships at each level of the other independent variable" (p. 27). In the model containing an interaction term, the coefficients represent "conditional" relationships: "b_1 reflects the influence of X_1 on Y when X_2 equals zero, and b_2 reflects the influence of X_2 on Y when X_1 equals zero" (Jaccard et al., 1990, p. 27).

The above material constitutes an introduction to the investigation of interaction effects in multiple regression, but it is only an introduction. We have not addressed models with three independent variables, though the extension to this case is straightforward. We also have not addressed interaction between one continuous and one dummy variable or between two dummy variables. In addition, we have not addressed the case of other, more complex functional forms of interaction, such as when the effect of an independent variable on a dependent variable is a quadratic function of a third variable. The discussion of these topics is far beyond the scope of this presentation. We recommend Cohen and Cohen (1983), Kerlinger and Pedhazur (1982), and, particularly, Jaccard et al. (1990) for more complete discussion of these and other topics in multiple regression analysis.

NOTES

1. Under certain conditions, which we will discuss in the following sections, the independent variables may be dummy coded discrete variables.

2. Some authors use the Greek letter beta, β, but we reject this usage because of the confusion with Type II error and the fact that Greek letters typically are used to refer to population parameters. Throughout this book, we consistently write out the word "beta" or refer to standardized regression coefficients, or use the term "beta weight."

CHAPTER 11

The Bigger Picture

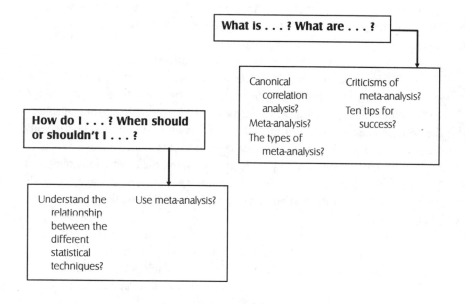

What is . . . ? What are . . . ?

- Canonical correlation analysis?
- Meta-analysis?
- The types of meta-analysis?
- Criticisms of meta-analysis?
- Ten tips for success?

How do I . . . ? When should or shouldn't I . . . ?

- Understand the relationship between the different statistical techniques?
- Use meta-analysis?

Level: Intermediate/Advanced

Focus: Conceptual/Instructional

The Bigger Picture

HOW DO I UNDERSTAND THE RELATIONSHIP
BETWEEN THE DIFFERENT STATISTICAL TECHNIQUES?

The way in which statistics typically is taught makes it likely that the beginning researcher will fail to appreciate the logical relationships among various statistical tests. Two group comparisons and multiple comparisons are often studied at different times, and different tests seem to emanate from different sources. Regression techniques and analysis of variance are a case in point. They do indeed derive from different research traditions, yet there is a logical coherency behind inferential statistics that belies the distinctions, and there are indeed mathematical commonalities among the basic methods that enable the researcher to derive one technique from another. We attempt to capture these commonalities in Figures 11.1 and 11.2. Figure 11.1 moves from more complex analyses (canonical correlation) with multiple independent and dependent variables, continuously distributed, to a less complex analysis (bivariate cross-tabulation), with one independent and one dependent variable, both discrete. Figure 11.2 begins with two continuously distributed variables, one independent and one dependent, and asks the question, "What techniques would we use if we increased the number of variables or if one or more of the variables became discrete instead of continuous?" Of course, these are simplifica-

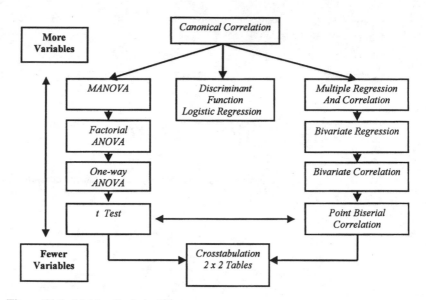

Figure 11.1. Linking Statistical Tests

tions, and there are many more techniques that could have been added; however, the goal is to show that the most commonly used techniques are logically linked into a framework. The framework that ties these techniques together is based on the number of independent and dependent variables and their level of measurement.

Figure 11.1 describes permutations of the general linear model. We begin our journey with **canonical correlation analysis (CCA)**, the most generic form of multivariate technique that relates a set of independent variables to a set of dependent variables (Knapp, 1978). Any number of independent and dependent variables qualify. Moreover, the independent variables, though typically considered continuous, could be dummy coded categorical variables. As we move down and to the right, we come to multiple regression. This is a special case of canonical correlation analysis where there is only one dependent variable and two or more independent variables, some or all of which could be dummy coded categorical variables. If the dependent variable is categorical, we could analyze the data with discriminant function analysis or logistic regression analysis, both shown in the middle of the diagram. The multiple correlation coefficient describes the relationship between the set of independent variables and the dependent variable. When there is only one independent variable, we have bivariate correlation, a special case of multiple regression, and when that dependent variable is dichotomous rather than continuous, the measure is called a point biserial correlation coefficient and correlates two groups of subjects (e.g., men and women), with a continuous variable (Baggaley, 1981). When both the independent variable and the dependent variable are dichotomous, the result is a 2×2 table, and the appropriate correlation is known as a phi coefficient (which is closely related to a chi-square value associated with the table). A phi coefficient is simply a correlation between two dummy variables, and thus is a special case of bivariate correlation.

The point biserial correlation can be converted in a straightforward manner into a t value, because a t test examines the relationship between a dummy variable (i.e., two independent groups) and a continuous variable. In fact, the significance test for r is identical to the significance test associated with the t test. Furthermore, the t test is a special case of a one-way analysis of variance for two independent groups of subjects (i.e., two levels of one factor). In fact, $F = t^2$ for independent groups. Moving further along the circle, we see that ANOVA is a special case of multivariate analysis of variance (MANOVA) in which there is only one dependent variable. Because MANOVA is a technique for analyzing data with categorical independent variables and continuous dependent variables, it can be seen as a special case of CCA. This completes the circle tour.

Note that the right side of the diagram contains correlational types of procedures that are particularly relevant for studying problems in which the independent variables are traits or attributes (i.e., continuously distributed variables). The left side of the diagram is more associated with the traditional experimental paradigm, where independent variables are comparative interventions or procedures (i.e., categorical variables). The middle of the diagram contains discriminant function analysis and logistic regression analysis, appropriate for use when the dependent variable is categorical. When there are more than two groups composing the dependent variable, these techniques are called multiple discriminant function analysis and polytomous logistic regression. We can think of these as special cases of multiple regression with categorical dependent variables. Discriminant function analysis and logistic regression can be seen as reversing the logic of ANOVA by examining the continuously distributed predictors of group differences rather than the ability of group differences to predict continuously distributed outcome variables. To a large extent, the appropriate choice of statistical technique is a function of the number and type of variables the researcher is interested in relating to one another. The conversion from one technique to the next is then a function of mathematical calisthenics, best performed by a computer.

Number of Independent Variables   **Number of Dependent Variables**

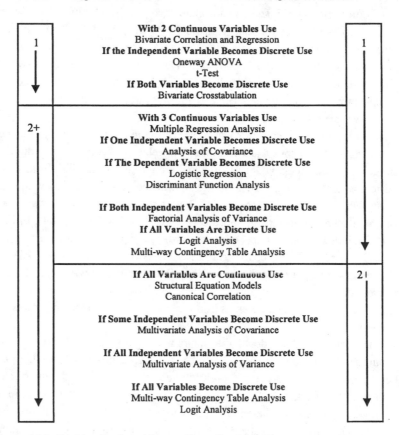

Figure 11.2. Linking Statistical Tests—From Simple to Complex Analyses

Finally, it should be evident that this diagram does not include a large number of more esoteric statistical tests, and it excludes some of the powerful, newer approaches to data analysis, such as path analysis and confirmatory factor analysis, as well as logit and probit analysis. It is also possible to demonstrate the relationship between the canonical correlation analysis model and these techniques; however, these techniques—and logistic regression, which we have included—are not typically considered part of the "general linear model."

Figure 11.2 does much the same thing as Figure 11.1, but in a slightly different way. In this figure, we begin with a bivariate correlation and ask the question, "What if?" For example, if the independent variable was discrete as opposed to continuous, what technique would be appropriate? The answer is, one-way analysis of variance, or a t test, in the case of a dichotomous independent variable. Again, we make the point that in order to understand the relationships among tests, and to select the appropriate test, we must know the number of independent and dependent variables and their level of measurement.

Researchers have been using the techniques shown in Figures 11.1 and 11.2 for 50 years. What has happened to all of this research? It would seem a good idea to organize all the research on a topic, mix it together, and ask the question, "What do we know?" This is what typically has become known as the "literature review." In the past few years, a particular type of literature review has become prominent. Known as "meta-analysis," it summarizes the quantitative knowledge in an area using a new set of statistical procedures that allows the researcher to gauge the overall magnitude of the effects shown in multiple studies. We consider these techniques in the next section.

WHEN SHOULD I USE META-ANALYSIS?

Meta-analysis is probably the most radical departure from significance testing seen in the social sciences in the past 50 years.[1] Most of us have been inculcated with the dominant orientation in statistical thinking, which assumes that knowledge in the social sciences rests on the accumulation of results from definitive individual empirical studies. In contrast, meta-analysis is a form of secondary analysis of pre-existing data that aims to summarize and compare results from different studies. Literally thousands of meta-analyses have been conducted, addressing many different literatures. These analyses serve to combine results from multiple studies and, consequently, allow us to diminish our reliance on statistical tests from individual studies. Adoption of the term *meta-analysis* generally is credited to Glass (1976), who did not intend to invent a new form of statistical analysis but proposed meta-analysis as an "analysis of analyses" using previously familiar methods that go back to the time of Fisher and Pearson. To be fair, there are numerous early precursors of meta-analysis within the category of integrative reviews of the literature in a particular area (cf. Ghiselli, 1949; Underwood, 1957). What makes the meta-analytic method unique is moving beyond summarizing a literature by counting or combining significance levels to incorporating the concepts of power and effect size to integrate the findings of diverse studies (Bangert-Drowns, 1986).

Meta-analysis has been described by Abelson (1995) as akin to having a research community who pool results across studies to provide a summary measure of the relationship between variables or the size of a phenomenon. Compare this perspective with researchers who magnify the importance of a single study because they generalize the results of the study to the population, believing that statistical significance tests evaluate the probability of population parameters (see Chapter 4 for an explanation of this controversy). Thus, meta-analysis serves as an antidote to both the shoddy accumulation of research knowledge and the recurrent presence of small effects. Because meta-analysis takes a number of directions, it is not surprising that several forms of meta-analysis exist. These forms have been summarized by Bangert-Drowns (1986) into five categories, described in the following paragraphs.

First, Glass and his colleagues (Glass, McGaw, & Smith, 1981) use liberal criteria to select all relevant studies on a particular research question. They then transform the results of the various studies into a common metric, the standard effect size, ES (Cohen's d, the difference between the experimental and control group means divided by the standard deviation of the control group), to determine an average outcome across the group of studies. The unit of analysis is a finding of the study, which means that each study can have several findings with several dependent variables, each generating a separate effect size.

Second, the Study Effect Meta Analysis (SEM) approach of Mansfield and Bussey (1977) uses the study, rather than the finding, as the unit of analysis. If a study has more than one dependent measure, the respective effect sizes are either combined if they reflect the same construct (e.g., physical pain) or analyzed separately if they reflect different constructs (e.g., physical pain and depression). The SEM approach is generally more selective in choosing studies for review, rejecting studies that do not meet carefully delineated methodological criteria.

Third is the Combined Probability Method recommended by Rosenthal and his colleagues (cf. Rosenthal & Rubin, 1986). In this approach, the focus is on estimating a treatment effect across studies and determining the reliability of the estimate, rather than reviewing a literature. The researcher addresses all relevant studies and calculates one-tailed p values and their corresponding standard normal deviate (Z) for each study, as well as Cohen's d, to generate an average effect size and combined probability measures.

Fourth, Approximate Data Pooling With Tests of Homogeneity is an extension of Rosenthal's Combined Probability Method. As opposed to critically reviewing a large set of studies, this approach relies on summary statistics from the studies to help pool all subjects across the studies into one large comparison. Hedges and Olkin (1985) and Rosenthal and Rubin (1982) have created tests to evaluate the homogeneity of effect sizes of the pooled studies that can guide the decision of how and if to create an average effect size.

Fifth, Approximate Data Pooling With Sampling Error Correction is a variation of the previous approach devised by Hunter and Schmidt (1990). Both approaches use a similar strategy: All studies of interest are represented by a single effect size and included regardless of methodological rigor. No tests of homogeneity of effect sizes are conducted, however. Rather, Hunter and Schmidt correct the effect sizes for unreliability and adjust their variability for sampling error.

The reader is referred to the previously cited references for more details about each of these approaches to meta-analysis. Each, of course, has its strengths and weaknesses. For instance, Glass has been criticized for not being sufficiently selective about the quality of individual studies that he chooses to include in his meta-analyses. Moreover, generating multiple effect size estimates for individual studies implies including non-independent data points in the meta-analysis. Rosenthal and Rubin (1986) provide specific recommendations to overcome this problem by computing summary effect sizes from studies based on multiple dependent variables. The reader is also referred to a recent article by Ray and Shadish (1996) that shows how using different measures of effect size in meta-analyses can yield somewhat different conclusions.

The five considerations outlined in the following paragraphs should inform any meta-analytic review.

First is deciding on the purpose of the review. As Bangert-Drowns (1986) reminds us, some approaches, such as that of Glass and his colleagues, aim to provide a summary of a research literature that addresses the relationship between two or more variables or the impact of a treatment on a group of participants. Other approaches test a particular hypothesis with a larger sample size than is available through a single study. These techniques attempt to determine a treatment effect across studies and, as such, are more analogous to methods of secondary analysis, such as cluster analysis, than to a narrative review of the literature.

Second is choosing studies for inclusion in the meta-analysis. This means conducting a search of the literature for studies meeting certain inclusion criteria. These criteria usually include the elimination of studies that do not meet *a priori* standards of substantive relevance and suitable methodological rigor. The inclusion of poorly designed studies can easily pollute ultimate conclusions, although one has to be careful not to base inclusion decisions on researcher bias. Certainly, the independent and dependent variables must be defined explicitly, so that the results can be evaluated and replicated. Once the criteria for selection have been established, several strategies may be considered for gathering the studies. These techniques include the use of computer databases, manual searches of relevant journals, and the ancestry, descendence, and invisible college approaches, all of which are described by Johnson (1989).

Third is determining the unit of analysis. Effect sizes can be calculated on the basis of one per study or one per finding within a study. In the event of multiple dependent variables, we generally recommend averaging effect sizes for each situation and using one effect size per study, so that no single study carries undue influence in the overall analysis. We also agree with Thompson (1997) that effect sizes should be described in an "accessible metric," in other words, in language that is clearly understandable and relevant to the topic. An example is number of working days lost due to stress, taken from studies on the impact of organizational stressors. The reader is referred to an article by Saunders, Howard, and Newman (1988) on how to express effect sizes meaningfully.

Fourth is choosing the specific effect size measure. This determination depends on the form of the data. Most meta-analyses express the effect size as Cohen's d or Hedges and Olkin's (1985) g, an unbiased estimator of the population d. This measure is derived from means and standard deviations within the data set. When these data are unavailable, estimates of d can be achieved from the available correlations, using the formula

$$d = \frac{2r}{\sqrt{1-r^2}}.$$

Fifth is computing a weighted mean effect size from the individual studies. The statistical significance of the effect size (d) can easily be determined from the variance and sampling error of the d values. The homogeneity or heterogeneity of the overall effect size can also be determined (see Hedges & Olkin, 1985). When effect sizes are heterogeneous, they do not derive from a common population, meaning that there are moderator variables present that differentiate the various studies. These moderators can be identified with focused contrast tests that check the influence of specific attributes that represent key differences among the studies. In addition to moderator analysis, it may also be prudent to conduct a diagnosis and analysis of outliers (see Chapter 7) and to display the data visually using stem-and-leaf diagrams or boxplots (see Chapter 2).

? The Pros and Cons of Meta-Analysis

Perhaps the most passionate statement in favor of meta-analysis has been voiced recently by Schmidt (1996). Schmidt's argument is based on viewing meta-analysis as an antidote to his disenchantment with traditional significance testing. The problems with statistical significance testing have been raised in Chapter 4. What is relevant in this context is the observation that in most surveys of a body of research on a particular research problem, some studies will have statistically significant findings and some will not. The general tendency is to honor those studies whose findings reach statistical significance and ignore those that do not, or, perhaps more disempowering, to be paralyzed in moving forward in the pursuit of knowledge because of the apparent contradictions across studies. It should be clear, argues Schmidt, that reliance on dichotomous rules for statistical significance is highly misleading: What is important is the size of the effect, not the size of the p level. Moreover, an increasing emphasis on the power of studies (minimizing Type II error) has meant that it is increasingly difficult for the individual researcher, forced to rely on relatively limited sample sizes, to achieve results that are statistically significant. This is especially the case when the true effect size is relatively small. Meta-analysis can easily include statistically nonsignificant findings to arrive at a truer picture of the robustness of an effect (although nonsignificant findings are not easily published!).

Meta-analysis is not without its critics. Below, we briefly discuss five of the most frequently voiced criticisms of the general meta-analytic approach.

First, a frequently heard criticism is that bringing together studies that may utilize different measures of the same variables, different statistical techniques, and different settings is like comparing apples and oranges. This criticism is lodged particularly against researchers like Glass (see Glass et al., 1981), who has been accused of relying on overly broad categories to summarize effects across variables. Cohen (1962), Dubin and Taveggia (1968), and others have countered this criticism by providing inventive solutions to compare diverse methodologies.

A second criticism is that the significance tests used in meta-analysis are often too liberal (Hedges, 1983). This is because average effect sizes might be artificially inflated as a result of, among other things, non-independence of data points, differences in measurement error among studies, or interactions of treatments with random contextual factors. Researchers do need to consider and correct for such statistical artifacts. Methods for doing so are available (cf. Schmidt & Hunter, 1996).

A third criticism has been voiced by those who believe that the widespread use of meta-analysis is an invitation to take a purely descriptive, empirical approach to research and avoid the more theory-based, conceptual work that goes into designing and conducting individual studies. We would respond that there is ample room for both perspectives and that either approach works at its best when accompanied by heavy doses of conceptual thinking and empirical precision.

A fourth criticism is directed at the loss of information that occurs through the methods of meta-analysis. In one regard, this can refer to the lack of representativeness of the studies selected for a meta-analysis from the entire pool of studies conducted on the same question. (Strictly speaking, to use inferential statistics on meta-analytically generated data, one should randomly sample the selected studies from the population.) In another way, this criticism concerns the loss of information that ensues from reducing a body of literature to a single measure such as a mean effect size. In response, Rosenthal (1995) reminds us that the attempt to understand differences among the results of diverse studies should be an inherent part of a meta-analysis.

A final criticism comes from those who worry that a preoccupation with meta-analysis will lead us to disregard those unique studies that offer something surprising and potentially important that has been revealed by traditional significance testing. The counterargument, of course, is that individual studies may indeed lead to the repeal of an old explanation or the development of a new perspective, but that without replications, they cannot be taken too seriously.

? An Example of Meta-Analysis

Providing a comprehensive example of meta-analysis is beyond the scope of this book. It is safe to say, however, that any meta-analysis includes two fundamental components: deriving an effect size estimate from each study in the analysis and using statistical procedures to aggregate these estimates across studies (Morris & DeShon, 1997). More effort has been spent on the second task, how to aggregate estimates from diverse studies (cf. Johnson, Mullen, & Salas, 1995; Law, Schmidt, & Hunter, 1994) than on the first issue, computing the effect size estimates. A major challenge is to provide comparable estimates from studies that rely on very different statistical methods and designs. The reader is referred to the writings of Dunlap, Cortina, Vaslow, and Burke (1996), Nouri and Greenberg (1995), and Morris and DeShon (1997) for recent contributions

TABLE 11.1 Ten Sample Correlations From the Same Population (population value = −.243)

Sample Number (p value)	N	r	95% Confidence Interval
1 (.054)	25	−.407	(−.748, −.066)
2 (.018)	25	−.480	(−.794, −.166)
3 (.487)	50	−.103	(−.386, +.180)
4 (.066)	50	−.268	(−.533, −.002)
5 (.047)	75	−.232	(−.452, −.012)
6 (.044)	75	−.235	(−.455, −.015)
7 (.018)	100	−.242	(−.430, −.053)
8 (.001)	100	−.395	(−.564, −.226)
9 (.003)	125	−.267	(−.434, −.100)
10 (.008)	125	−.217	(−.388, −.046)

to solving this problem. Below, we provide a sample analysis based on correlation coefficients.

To illustrate a rudimentary meta-analysis, we sampled 10 correlations from a population with a known correlation of −.243. The number of cases in the population was approximately 1,000 and we drew 10 samples, two each of 25, 50, 75, 100, and 125 cases. The results are presented in Table 11.1.

The correlations range in size from −.480 to −.103. Five of the samples overestimated and five underestimated the population value. The computer program we used (SPSS® for Windows 95®) indicated that samples 1, 3, and 4 were not statistically significant at the .05 alpha level, as shown by the p values in parentheses following the sample number. We next calculated the 95% confidence interval using a t value of 2 and the formula $(1 − r^2)/\sqrt{N − 1}$ for the standard error. Of the 10 confidence intervals, only the third included the value of zero. (The p values for the 1st, 3rd, and 4th correlations were given as .054, .487, and .066, respectively. The discrepancy is due to our method of calculating the confidence intervals, as the 1st and 3rd confidence intervals should include the value of zero to be totally consistent with the computer-based p values.) The reader should note that *all* the confidence intervals include the population parameter, but for the smaller samples, the confidence intervals indicate that the correlation might be as large (in absolute value) as −.794, or as small as −.002, or even positive (.18, sample 3). In addition, if any of these smaller sample sizes

TABLE 11.2 Meta-Analysis of Ten Samples From the Same Population

Number of Samples	Mean	95% Confidence Interval
10	−.284	(−.364, −.205)

(50 or less) actually obtained the correct population value (i.e., −.243), the null hypothesis would not be rejected at the .05 alpha level. In the first four samples that we drew, with sample sizes of 25 and 50 cases, three of the four sample values were larger (in absolute value) than the population value, and three of these were not reported as statistically significant at the .05 alpha level. The traditional conclusion for these outcomes would be that no relationship exists. This is a conclusion that we know is false and leads to a Type II error. The larger samples tend to provide more precise estimates of the population value because the 95% confidence interval represents a narrower range of values. Nevertheless, even the closest estimate (sample 7), based on 100 cases and a having value of −.242, indicates a possible range for the population correlation between −.430 and −.053. Substantively, this means that the relationship could be quite strong and potentially very important, or very weak and substantively meaningless.

We could use a number of strategies to conduct our meta-analysis on these data. One would be to calculate the d value, using the formula shown earlier, and then determine the expected variability around these values. (Formulas are provided in Hunter & Schmidt, 1990.) We simply calculated the mean and 95% confidence interval for the 10 sample correlations. The results are presented in Table 11.2.

The mean of the 10 correlations is −.284, and the 95% confidence interval is −.364 to −.205. Note that this gives a better estimate of the population value than most of our single samples, particularly the ones based on small sample sizes, and narrows the interval within which we expect the population value to fall. Although there certainly are other ways we could manipulate these data to provide meta-analytic estimates of the population parameter, this example illustrates the basic logic behind the meta-analytic approach. First, no significance tests are used; and second, effect sizes, represented by the r values, are the central focus of the investigation. Although this example combined samples from the same population, it illustrates the range of values, and the different conclusions, that might be reached based on random samples of different sizes.

```
.5 | 2 5
.4 |
.3 | 0
.2 | 1 7
.1 | 4 6 8
.0 | 2 5
-.0 | 6 4 3 2 2 1
-.1 | 3 3
```

Figure 11.3. Stem-and-Leaf Diagram of 18 Studies of the Pygmalion Effect
SOURCE: Adapted from Raudenbush (1984).

The fact that all these confidence intervals overlap to some extent correctly implies substantial agreement among the studies, even though traditional significance testing might lead to the conclusion that some of the studies show no relationship between the values and some would lead to the opposite conclusion. Thus, meta-analysis helps us elaborate on the meaning of individual studies and draw conclusions about accumulated research findings. In this era of information overload, meta-analysis helps us communicate succinctly about the cumulative conclusions of a body of research by generating a single population value. With multiple samples from many different studies of the same phenomenon, meta-analysis appears likely to facilitate the realization of more reasonable conclusions regarding population effect sizes.

Finally, it is possible to use meta-analysis to explore the impact of contextual variables on primary relationships between independent and dependent variables (Abelson, 1995). Abelson cites a meta-analysis of Pygmalion effects on IQ scores from data adapted from Raudenbush (1984). The Pygmalion effect refers to the self-fulfilling prophecy imposed by classroom teachers who mistakenly have been led to believe that particular students, actually chosen randomly, are likely to bloom academically in comparison to their peers (Rosenthal & Jacobson, 1968). Figure 11.3 is a stem-and-leaf summary of the effect sizes of 18 studies on the Pygmalion effect reviewed by Raudenbush (1984).

Each number refers to the mean difference between the IQ scores of bloomers and controls, divided by the pooled within-group standard deviation. The minus signs refer to the eight studies with results in the unanticipated direction. The mean effect size of the 18 studies is a modest .109, which is statistically significant but refers to an IQ gain of only about 1.5 points. Raudenbush carried the analysis one step further, however, and tried to account for the fact that a few studies had larger effect sizes (more than .20, representing IQ gains of 3 to 8 points). What he noted was that in four studies, the teacher had never met the students prior to the study (those with effect sizes in bold numbering), and in three studies, the teacher had known the students only for a week (those with effect sizes in italics). Six of these seven cases represent the highest effect sizes among the studies, suggesting that prior exposure to the students is a mediating variable in the analysis.

It would have been easy to miss this wrinkle on the Pygmalion effect without the aid of meta-analysis. The procedure has the advantage of allowing for the simultaneous inspection of several contextual variables in interaction with the treatment effect. As we know, potential mediating or moderating variables can have effects even if they have never been identified through experimental manipulation. As Abelson (1995) puts it, "You can't see the dust if you don't move the couch" (p. xv).

TEN TIPS FOR SUCCESS
IN STATISTICAL ANALYSIS

We have now presented some ways to think about the planning of statistical analyses and made several recommendations about managing challenging or confusing issues in this field. By way of summary, here are 10 rules of the road, so to speak, that we believe deserve attention in regard to summarizing, discussing, and making decisions about statistical analyses.

? 1. Get Comfortable With Your Data!

It is extremely important to familiarize yourself thoroughly with your data prior to conducting analyses that will almost undoubtedly hide the basic structure of the data. This is particularly true when conducting secondary analyses or when someone else has built the database with which you are working. You should know, for example, which variables are continuous and which are categorical. You should know how dummy variables have been coded: Is it 0, 1 or 1, 2? You should know what the high and low ends of a scale mean. For example, in assessing locus of control, does a high score mean internal locus of control or external locus of control? Or, perhaps, was the locus of control variable coded as a dichotomy? Does "1" mean internal and "0" external, or is "1" external and "0" internal? You become familiar with your data by working with them and having participated in their collection and condensation into a database. We believe this is necessary to avoid making critical mistakes. Why do we mention something that seems so simple and obvious? Because we have seen numerous instances where individuals seeking our help and advice cannot answer even the most basic questions about their own data.

? 2. Thoroughly Explore Your Data, Twice!

In line with the first suggestion, we recommend that you look at your data, using both graphs and tables, prior to analyzing them with statistical techniques that may obscure, rather than elucidate, the true character of the data. Previously in this book, we introduced the use of stem-and-leaf diagrams and box and whisker plots in addition to traditional histograms as ways of capturing statistical distributions. We wish once again to highlight the importance of exploratory data analysis at both the beginning and the end of the data analysis process. At the beginning, it is important to explore the data to understand the nature of the distributions of the variables. Do the data make sense in terms of the range of legitimate values? Are the categorical variables evenly distributed, or are there categories containing a very large or very small percentage of the sample? Are continuous variables normally distributed or skewed? How do the data distributions characterize the sample?

It is also important to explore the data at the end of the data analysis process. It is not uncommon to find that researchers have missed an opportunity to explore their data using alternative analytic perspectives that might support their theoretical framework, or suggest new patterns of relationships that should be explored further. Such explorations can involve the use of graphic techniques (described above), different statistical methods, or a combination of both. We do not mean that you should sift your data in every possible way to find some statistically significant finding just to highlight it and disregard all the remaining results. We do mean, however, that there is nothing wrong with considering alternative hypotheses and alternative relationships from your data set. There is also nothing wrong with analyzing your data in more than one way. After all, you have likely spent a great deal of time and effort designing and conducting your study. Why not give equal consideration to the meaning of the data? It is not at all unusual for the most interesting findings to exist where you least expect them. That is serendipity, and that is science!

Researchers can become "stuck" with one favorite method or analytic strategy that characterizes an area of research and that blinds them to fruitful alternative approaches. A word of warning, however, is in order. Mining your data is not the same thing as post hoc theorizing. Most good research is theory driven, meaning research questions and research designs are a function of a logical thinking process guided by existent theory and research. Explaining results after the fact or statistically testing alternative hypotheses without due consideration of the implications of sampling error is not recommended. Such new discoveries should lead to the initiation of additional studies, which brings us to our next recommendation.

? 3. Sometimes Pictures Speak Louder Than Words

We have emphasized the importance of graphic displays in understanding the meaning of data. A visual representation of your data may disclose the possible meaning and implications of your study in a way that abstract numbers may conceal. After analyzing your data using the appropriate descriptive and inferential statistics, consider accompanying your results with appropriate tables and graphs. Most statistical programs, as well as database and spreadsheet programs, have the capacity to produce attractive graphs very quickly. Such presentations can reveal relationships among variables that would remain hidden with tabled data. There are many good sources for learning about the appropriate use of graphs with scientific data (cf. Cleveland & McGill, 1985; Wainer, 1984). When writing research articles for publication, you must ask yourself if the graph provides additional information above and beyond tabled data. Then, too, some graphs appear so complicated that the average reader is unable to make sense of the data the researcher is attempting to simplify! People have a hard time, for instance, comprehending the visual presentation of three-dimensional space (and certainly, more than three dimensions!). Try not to be seduced by the easy availability of graphic presentation in most statistical programs. There is no need and no advantage in presenting data in a graph that can be easily and succinctly summarized in a table, unless the graph shows something special.

? 4. Replication Is Underemphasized and Overdue

Statistically significant results need to be replicated with new samples and new settings. The research world is much too full of isolated studies that yielded significant results with idiosyncratic samples under particular circumstances. Most of these studies have never been replicated and should be regarded as untrustworthy in terms of generalizing their findings to a larger context. One of the primary reasons for the recent popularity of meta-analysis is the opportunity to pull together diverse studies on common themes in an effort to draw conclusions that have significant import in terms of application and theory building. In science, something worth doing is worth doing over and over again. Replicating a good study or improving on a flawed study is often a better choice than adding another trivial study to the literature.

? 5. Remember the Distinction Between Statistical Significance and Substantive Significance

Most social scientists have been seduced by the sanctity of a statistically significant p value. As you now know, there is a sea change taking place within the world of statistics, questioning the logic behind statistical decision-making theory. At the very least, we know that there is nothing sacred about a particular p value. To a large extent, statistical significance is a function of sample size. Just because a finding is statistically significant does not mean that it is of theoretical or practical interest. Likewise, a study that does not reach statistical significance is not necessarily a worthless study.

One important trend is to provide a specific probability level rather than merely to indicate whether or not the test value is significant beyond a certain traditional level, such as $p < .05$. This allows the reader the freedom to determine if the data reject or support the research hypothesis. Moreover, attention needs to be paid to substantive significance.

? 6. Remember the Distinction Between Statistical Significance and Effect Size

We have seen numerous published studies that focus only on significance levels. Try to approach every study as addressing *both* the question of statistical significance vs. substance significance (Suggestion 5, above) and the question of statistical significance versus size of effect. Both of these must be taken into consideration to fully judge the substantive significance of a research study. Effect sizes and confidence intervals should find a place in your analyses and presentations of results.

One suggestion is to present a confidence interval around an effect size measure, such as a correlation or standardized mean difference. If the interval includes the value of zero (or some other value indicating lack of support for a research question), then we know that the finding was not statistically significant at the specified level of confidence. We also learn, however, the range of potential effect sizes in the population, a piece of valuable information we almost never get in studies focusing only on statistical significance.

? 7. Statistics Do Not Speak for Themselves

The goal of statistical analysis is not simply to produce a bunch of tables and graphs. Virtually any person with a minimum of training can accomplish that. The goal is to use statistics to present an organized argument that supports a particular position and/or fails to support alternative positions. It is the researcher's responsibility to lead the reader through analyses that often do not speak for themselves. The most common pitfalls are to present too much analysis with too little explanation, or too little information with overattention to detail and little or no attention to the overall meaning of an analysis. A responsible researcher leads the reader through the findings, pointing out those portions that support his or her argument and those that fail to do so. Then the researcher should tie the findings to the theoretical and/or applied questions that generated the research in the first place, including the relevant literature and the possible outcomes of alternative designs.

? 8. Keep It Simple When Possible

Complex statistics sometimes are used to cover up a lack of understanding of a phenomenon and, ironically, can lead to greater confusion. We recommend choosing the most elementary, straightforward design and analysis that do justice to your research questions. There obviously is a place for sophisticated analyses, but simple techniques are easier and less prone to error in administration, interpretation, and the communication of results.

? 9. Use Consultants

The field of statistics can be difficult, confusing, and incredibly frustrating. It is perfectly acceptable and highly recommended that one consult experts. Surprisingly enough, it is often statisticians themselves who are more comfortable with consulting experts. Perhaps this is because they recognize the complexity and vastness of the field. No one person will be an expert in all areas, and more typically, statisticians consider themselves "really comfortable" with only a few techniques. Find the people who can help you, and make use of their advice. In line with this advice, it is *not* the math department or computer science department of a university where the most appropriate persons typically are found. For social science research, we recommend finding a sociologist, psychologist, or political scientist who is knowledgeable, who is willing to explain things slowly and clearly, and who speaks English, not "statspeak."

? 10. Don't Be Too Hard on Yourself

Our consulting experience indicates that few studies support all the hypotheses set forth at the beginning and that the effects are almost always weaker than expected. This is because as researchers, we become excited about the power of theoretical arguments and committed to demonstrating their truth. In reality, even findings that are both strong and consistent only add support to a body of research, suggesting that there may be some validity to a theoretical approach to social processes or behavior. When research findings are marginal, researchers often become disappointed and disillusioned, thinking that their work is a failure. Often, this is far from the truth, and the research has in fact made a meaningful contribution to understanding the phenomenon in question. The fault lies in expectations that are not tempered with a lengthy background in the reality of social science research. Measures are unreliable, samples often are not random, sample sizes often are too small, and the theories that we test are vague and incomplete. The researcher has a mound of difficulties to overcome, and good research can be difficult, expensive, and time-consuming. Our message is to lower your expectations and not reject an entire study because only some of the findings produce meaningful results. Look for the overall pattern of results—significant and insignificant, weak and strong—fit these into the original questions, and use them to generate new ideas and subsequent studies.

NOTE

---

1. Several basic texts (cf. Glass, McGaw, & Smith, 1981; Hedges & Olkin, 1985; Hunter & Schmidt, 1990) provide thorough introductions to this increasingly popular topic.

References

Abelson, R. P. (1985). A variance explanation paradox: When a little is a lot. *Psychological Bulletin, 97,* 129-133.

Abelson, R. P. (1995). *Statistics as principled argument.* Hillsdale, NJ: Lawrence Erlbaum.

Axinn, S. (1966). Fallacy of the single risk. *Philosophy of Science, 33,* 154-162.

Babbie, E. (1997). *The practice of social research* (8th ed.). Belmont, CA: Wadsworth.

Baggaley, A. R. (1981). Multivariate analysis: An introduction for consumers of behavioral research. *Evaluation Review, 5,* 123-131.

Bakan, D. (1966). The test of significance in psychological research. *Psychological Bulletin, 66,* 423-437.

Bangert-Drowns, R. L. (1986). Review of developments in meta-analytic method. *Psychological Bulletin, 99*(3), 388-399.

Becker, G. (1991). Alternative methods of reporting research results. *American Psychologist, 46,* 654-655.

Behrens, J. (1997). Principles and procedures of exploratory data analysis. *Psychological Methods, 2,* 131-160.

Berg, B. L. (1998). *Qualitative research methods for the social sciences* (3rd ed.). Boston: Allyn & Bacon.

Bird, K. D., & Hadzi-Pavlovic, D. (1983). Simultaneous test procedures and the choice of a test statistic in MANOVA. *Psychological Bulletin, 93,* 167-178.

Blalock, H. M. (1970). *Causal models in the social sciences.* Chicago: Aldine.

Blalock, H. M. (1979). *Social statistics.* New York: McGraw-Hill.

Bohrnstedt, G. W., & Knoke, D. (1994). *Statistics for social data analysis* (3rd ed.). Itasca, IL: F. E. Peacock.

Boneau, C. A. (1960). The effects of violation of assumptions underlying the *t* test. *Psychological Bulletin, 38,* 49-64.

Borgatta, E. F., & Bohrnstedt, G. W. (1981). Level of measurement: Once over again. In G. W. Bohrnstedt & E. F. Borgatta (Eds.), *Social measurement: Current issues* (pp. 23-27). Beverly Hills, CA: Sage.

Box, G. E. P., & Cox, D. R. (1964). An analysis of transformations. *Journal of the Royal Statistical Society, 26,* 211-243.

Bray, J. H., & Maxwell, S. E. (1982). Analyzing and interpreting significant MANOVAs. *Review of Educational Research, 52*, 340-367.

Campbell, D. R., & Stanley, J. C. (1966). *Experimental and quasi-experimental designs for research.* New York: Rand McNally.

Carmines, E. G., & McIver, J. P. (1981). Analyzing models with unobserved variables: Analysis of covariance structures. In G. W. Bohrnstedt & E. F. Borgatta (Eds.), *Social measurement: Current issues* (pp. 65-115). Beverly Hills, CA: Sage.

Carmines, E. C., & Zeller, R. A. (1979). *Reliability and validity assessment* (Sage University Paper series on Quantitative Applications in the Social Sciences, 07-017). Beverly Hills, CA: Sage.

Cattin, P. (1980). Note on the estimation of the squared cross-validated multiple correlation of a regression model. *Psychological Bulletin, 87*(1), 63-65.

Chow, S. L. (1988). Significance test or effect size? *Psychological Bulletin, 103*, 105-110.

Cleveland, W. S., & McGill, R. (1985). Graphical perception and graphical methods for analyzing scientific data. *Science, 229*, 828-833.

Cobb, E. B. (1985). Planning research studies: An alternative to power analysis. *Nursing Research, 34*, 386-388.

Cohen, J. (1962). The statistical power of abnormal-social psychological research: A review. *Journal of Abnormal and Social Psychology, 65*, 145-153.

Cohen, J. (1965). Some statistical issues in psychological research. In B. B. Wolman (Ed.), *Handbook of clinical psychology* (pp. 95-121). New York: McGraw-Hill.

Cohen, J. (1977). *Statistical power analysis for the behavioral sciences* (rev. ed.). New York: Academic Press.

Cohen, J. (1983). The cost of dichotomization. *Applied Psychological Measurement, 7*, 249-253.

Cohen, J. (1988). *Statistical power analysis for the behavioral sciences* (2nd ed.). Hillsdale, NJ: Lawrence Erlbaum.

Cohen, J. (1990). Things I have learned (so far). *American Psychologist, 45*(12), 1304-1312.

Cohen, J. (1994). The Earth is round (*p* < .05). *American Psychologist, 49*, 997-1003.

Cohen, J., & Cohen, P. (1983). *Applied multiple regression/correlation analysis for the behavioral sciences.* Hillsdale, NJ: Lawrence Erlbaum.

Cole, D. A., Maxwell, S. E., Arvey, R., & Salas, E. (1994). How the power of MANOVA can both increase and decrease as a function of the intercorrelations among the dependent variables. *Psychological Bulletin, 115*(3), 465-474.

Cook, R. D., & Weisberg, S. (1982). *Residuals and influence in regression.* London: Chapman & Hall.

Cook, R. D., & Weisberg, S. (1994). *An introduction to regression graphics.* New York: John Wiley & Sons.

Cortina, J. M., & Dunlap, W. P. (1997). On the logic and purpose of significance testing. *Psychological Methods, 2*(2), 161-172.

Cowles, M. (1989). *Statistics in psychology: An historical perspective.* Hillsdale, NJ: Lawrence Erlbaum.

Cramer, E. M., & Bock, R. D. (1966). Multivariate analysis. *Review of Educational Research, 36*, 604-617.

Crane, P. (1988). *Childbearing postponement: A comparative analysis of voluntarily childless wives and wives who intend to have children.* Unpublished doctoral dissertation, University of Southern California.

Crask, M. R., & Perrault, W. D., Jr. (1977). Validation of discriminant analysis in marketing research. *Journal of Marketing Research, 14*, 60-68.

Cronbach, L. J. (1957). The two disciplines of scientific psychology. *American Psychologist, 12*, 671-684.

Cronbach, L. J., & Furby, L. (1970). How we should measure "change"—or should we? *Psychological Bulletin, 74*, 68-80.

Crosbie, J. (1993). Interrupted time-series analysis with brief single-subject data. *Journal of Consulting and Clinical Psychology, 61*(6), 966-974.

Dar, R., Serlin, R. C., & Omer, H. (1994). Misuse of statistical tests in three decades of psychotherapy research. *Journal of Consulting and Clinical Psychology, 62*(1), 75-82.

Darlington, R. B. (1968). Multiple regression in psychological research and practice. *Psychological Bulletin, 69*, 161-182.

Darlington, R. B. (1990). *Regression and linear models.* New York: McGraw-Hill.

Diaconis, P., & Efron, B. (1983). Computer-intensive methods in statistics. *Scientific American, 248*(5), 116-130.

Dubin, R., & Taveggia, T. C. (1968). *The teaching-learning paradox: A comparative analysis of college teaching methods.* Eugene: University of Oregon Press.

Dunlap, W. P., Cortina, J. M., Vaslow, J. B., & Burke, M. J. (1996). Meta-analysis of experiments with matched groups or repeated measures designs. *Psychological Methods, 1*(2), 170-177.

Einot, I., & Gabriel, K. R. (1975). A study of the powers of several methods of multiple comparisons. *Journal of the American Statistical Association, 70*, 574-583.

Emerson, J., & Strenio, J. (1983). Boxplots and batch comparison. In D. Hoaglin, F. Mosteller, & J. Tukey (Eds.), *Understanding robust and exploratory data analysis* (pp. 58-87). New York: Wiley.

Fisher, R. A. (1925). *Statistical methods for research workers.* Edinburgh: Oliver and Boyd.

Fisher, R. A. (1935). *The design of experiments.* Edinburgh: Oliver and Boyd.

Fisher, R. A., & MacKenzie, W. A. (1923). Studies in crop variation. II. The manurial response of different potato varieties. *Journal of Agricultural Science, 13*, 311-320.

Fox, J. (1991). *Regression diagnostics.* Newbury Park, CA: Sage.

Frick, R. W. (1996). The appropriate use of null hypothesis testing. *Psychological Methods, 1*, 379-390.

Galton, F. (1889). *Natural inheritance.* London: Macmillan.

Ghiselli, E. E. (1949). The validity of commonly employed occupational tests. *University of California Publications in Psychology, 5*, 253-288.

Gigerenzer, G. (1993). The superego, the ego, and the id in statistical reasoning. In G. Keren & C. Lewis (Eds.), *A handbook for data analysis in the behavioral sciences: Methodological issues.* Hillsdale, NJ: Lawrence Erlbaum.

Glass, G. V. (1976). Primary, secondary, and meta-analysis of research. *Educational Researcher, 5*(10), 3-8.

Glass, G. V., McGaw, B., & Smith, M. L. (1981). *Meta-analysis in social research.* Newbury Park, CA: Sage.

Gnanedesikan, R. (1977). *Methods for statistical analysis of multivariate observations.* New York: Wiley.

Godfrey, K. (1985). Comparing the means of several groups. *The New England Journal of Medicine, 313*(23), 1450-1456.

Gold, D. (1969). Statistical tests and substantive significance. *American Sociologist, 4*, 42-46.

Gordon, I. (1987). Sample size estimation in occupational mortality studies with use of confidence interval theory. *American Journal of Epidemiology, 125*, 158-162.

Gottman, J. M. (1981). *Time-series analysis: A comprehensive introduction for social scientists.* Cambridge, UK: Cambridge University Press.

Gottman, J. M., & Rushe, R. H. (1993). The analysis of change: Issues, fallacies, and new ideas. *Journal of Consulting and Clinical Psychology, 61*(6), 907-910.

Haase, R. F., & Ellis, M. V. (1987). Multivariate analysis of variance. *Journal of Counseling Psychology, 34*(4), 404-413.

Hair, J. F., Anderson, R. E., Tatham, R. L., & Black, W. C. (1995). *Multivariate data analysis* (4th ed.). Englewood Cliffs, NJ: Prentice Hall.

Hamilton, L. C. (1990). *Modern data analysis: A first course in applied statistics.* Belmont, CA: Wadsworth.

Hamilton, L. C. (1992). *Regression with graphics: A second course in applied statistics.* Monterey, CA. Brooks/Cole.

Hand, D. J., & Taylor, C. C. (1987). *Multivariate analysis of variance and repeated measures: A practical approach for behavioral scientists.* London: Chapman & Hall.

Harris, R. J. (1997). Significance tests have their place. *Psychological Science, 8*(1), 8-11.

Harwell, M. R. (1988). Choosing between parametric and nonparametric tests. *Journal of Counseling and Development, 67*, 35-38.

Hays, W. (1993). *Statistics* (4th ed.). New York: Holt, Rinehart, & Winston.

Hedges, L. V. (1983). A random effects model for effect size. *Psychological Bulletin, 93*, 388-395.

Hedges, L. V., & Olkin, I. (1985). *Statistical methods for meta-analysis.* New York: Academic Press.

Hendrick, C. (1990). Replications, strict replications, and conceptual replications: Are they important? *Journal of Social Behavior and Personality, 5*(4), 41-49.

Hertel, B. R. (1976). Minimizing error variance introduced by missing data routines in survey analysis. *Sociological Methods & Research, 4*, 459-474.

Hoaglin, D., Mosteller, F., & Tukey, J. (Eds.). (1983). *Understanding robust and exploratory data analysis.* New York: Wiley.

Hosmer, D. W., & Lemeshow, S. (1989). *Applied logistic regression.* New York: John Wiley & Sons.

Howell, D. C. (1997). *Statistical methods for psychology* (4th ed). Boston: Duxbury.

Huberty, C. J., & Morris, J. D. (1989). Multivariate analysis versus multiple univariate analyses. *Psychological Bulletin, 105*(2), 302-308.

Huck, S. W., & McLean, R. A. (1975). Using a repeated measures ANOVA to analyze the data from a pretest-posttest design: A potentially confusing task. *Psychological Bulletin, 82*(4), 511-518.

Humphreys, L. G. (1978). Research on individual differences requires correlational analysis, not ANOVA. *Intelligence, 2*, 1-5.

Humphreys, L. G., & Fleishman, A. I. (1974). Pseudo-orthogonal and other analysis of variance designs involving individual difference variables. *Journal of Educational Psychology, 66*, 464-472.

Hunter, J. E. (1997). Needed: A ban on the significance test. *Psychological Science, 8*(1), 3-7.

Hunter, J. E., & Schmidt, F. L. (1990). *Methods of meta-analysis.* Newbury Park, CA: Sage.

Jaccard, J., Becker, M. A., & Wood, G. (1984). Pairwise multiple comparison procedures: A review. *Psychological Bulletin, 96*, 589-596.

Jaccard, J., Turrisi, R., & Choi, W. K. (1990). *Interaction effects in multiple regression* (Sage University Paper series on Quantitative Applications in the Social Sciences No. 72). Newbury Park, CA: Sage.

Jacobson, N. S., Follette, W. C., & Revenstorf, D. (1984). Psychotherapy outcome research: Methods for reporting variability and evaluating clinical significance. *Behavior Therapy, 15*, 336-352.

Jacobson, P. (1976). *Introduction to statistical measures for the social and behavioral sciences.* New York: Holt, Rinehart & Winston.

Jeffreys, H. (1939). Random and systematic arrangements. *Biometrika, 31*, 1-8.

Jeffreys, R. C. (1990). *The logic of decisions* (2nd ed.). Chicago: University of Chicago Press.

Johnson, B. T. (1989). The effects of involvement, argument strength, and topic knowledge on persuasion. *Dissertation Abstracts International, 49*(10-B), 4604.

Johnson, B. T., Mullen, B., & Salas, E. (1995). Comparison of three major meta-analytic approaches. *Journal of Applied Psychology, 80*(1), 94-106.

Kachigan, S. K. (1986). *Statistical analysis: An interdisciplinary introduction to univariate and multivariate methods.* New York: Radius.

Kachigan, S. K. (1991). *Multivariate statistical analysis: A conceptual introduction* (2nd ed.). New York: Radius.

Kaiser, H. F. (1960). Directional statistical decisions. *Psychological Review, 67*, 160-167.

Kanji, G. K. (1993). *100 statistical tests.* Newbury Park, CA: Sage.

Kaplan, R. M., & Litrownik, A. J. (1977). Some statistical methods for the assessment of multiple outcome criteria in behavioral research. *Behavior Therapy, 8*, 383-392.

Kendall, M. G. (1962). *Rank correlation methods* (3rd ed.). London: Griffin.

Keppel, G. (1991). *Design and analysis: A researcher's handbook* (3rd ed.). Englewood Cliffs, NJ: Prentice Hall.

Kerlinger, F. (1986). *Foundations of behavioral research* (3rd ed.). New York: Holt, Rinehart & Winston.

Kerlinger, F., & Pedhazur, E. (1973). *Multiple regression in behavioral research.* New York: Holt, Rinehart & Winston.

Kerlinger, F. N., & Pedhazur, E. J. (1982). *Multiple regression in behavioral research: Explanation and prediction.* New York: Holt, Rinehart & Winston.

Kiecolt, J., & Nathan, L. (1985). *Secondary analysis of survey data* (Sage University Paper series on Quantitative Applications in the Social Sciences No. 53). Beverly Hills, CA: Sage.

Kirk, R. E. (1995). *Experimental design: Procedures for the behavioral sciences* (3rd ed.). Monterey, CA: Brooks/Cole.

Kirk, R. E. (1996). Practical significance: A concept whose time has come. *Educational and Psychological Measurement, 56*(5), 746-759.

Kish, L. (1995). *Survey sampling.* New York: Wiley. (Original work published 1965)

Knapp, T. R. (1978). Canonical correlation analysis: A general parametric significance testing system. *Psychological Bulletin, 85*, 410-416.

Kraemer, H. C., & Thiemann, S. (1987). *How many subjects? Statistical power analysis in research.* Newbury Park, CA: Sage.

Labovitz, S. (1970). The non-utility of significance tests: The significance of tests of significance reconsidered. *Pacific Sociological Review* (Summer), 141-148.

Law, K. S., Schmidt, F. L., & Hunter, J. E. (1994). Nonlinearity of range corrections in meta-analyses: Test of an improved procedure. *Journal of Applied Psychology, 79*(3), 425-438.

Lee, P. M. (1989). *Bayesian statistics: An introduction.* New York: Oxford University Press.

Lipsey, M. W. (1990). *Design sensitivity: Statistical power for experimental research.* Newbury Park, CA: Sage.

Lipsey, M. W., & Wilson, D. B. (1993). The efficacy of psychological, educational, and behavioral treatment. *American Psychologist, 48,* 1181-1209.

Little, R. J. A., & Rubin, D. B. (1987). *Statistical analysis with missing data.* New York: John Wiley & Sons.

Loftus, G. R. (1991). On the tyranny of hypothesis testing in the social sciences. *Contemporary Psychology, 36,* 102-105.

Loftus, G. R. (1997). Psychology will be a much better science when we change the way we analyze data. *Current Directions in Psychological Science, 5*(6), 161-171.

Loftus, G. R., & Masson, M. E. J. (1994). Using confidence intervals in within-subjects designs. *Psychonomic Bulletin & Review, 1,* 476-490.

Lord, F. M. (1967). Elementary models for measuring change. In C. W. Harris (Ed.), *Problems in measuring change* (pp. 199-211). Madison: University of Wisconsin Press.

Lunneborg, C. E. (1987). *Bootstrap applications for the behavioral sciences.* Seattle: University of Washington Press.

Lunneborg, C. E. (1990). Review of "Computer intensive methods for testing hypotheses." *Educational and Psychological Measurement, 50,* 441-445.

Magidson, J. (1979). *Advances in factor analysis and structural equation models.* London: University Press of America.

Mansfield, R. S., & Bussey, T. V. (1977). Meta-analysis of research: A rejoinder to Glass. *Educational Researcher, 6,* 3.

Maxwell, S., Delaney, H., & Dill, C. (1984). Another look at ANOVA versus blocking. *Psychological Bulletin, 95,* 136-147.

McGuire, W. J. (1989). A perspectivist approach to the strategic planning of programmatic scientific research. In W. R. Gholson, W. R. Shadish, Jr., R. A. Neimeyer, & A. C. Houts (Eds.), *Psychology of science: Contributions to meta-science* (pp. 214-245). Cambridge, UK: Cambridge University Press.

McMain, D. (1993). *McMain Marital Growth Inventory: Construction and validation of the instrument.* Unpublished doctoral dissertation, The Fielding Institute, Santa Barbara, CA.

Meehl, P. E. (1967). Theory testing in psychology and physics: A methodological paradox. *Philosophy of Science, 34,* 103-115.

Morris, S. B., & DeShon, R. P. (1997). Correcting effect size computed from factorial analysis of variance for use in meta-analysis. *Psychological Methods, 2*(2), 192-199.

Morrison, D. E., & Henkel, R. E. (Eds.). (1970). *The significance test controversy.* Chicago: Aldine.

Mosteller, F., & Tukey, J. W. (1977). *Data analysis and regression.* Reading, MA: Addison-Wesley.

Myers, J. L. (1979). *Fundamentals of experimental design* (4th ed.). Boston: Allyn & Bacon.

Nachmias, D., & Nachmias, C. (1995). *Research methods in the social sciences* (3rd ed.). New York: St. Martin's.

Neyman, J., & Pearson, E. S. (1933). On the testing of statistical hypotheses in relation to probabilities a priori. *Proceedings of the Cambridge Philosophical Society, 29,* 492-510.

Nouri, H., & Greenberg, R. H. (1995). Meta-analytic procedures for estimation of effect sizes in experiments using complex analysis of variance. *Journal of Management, 21*(4), 801-812.

Oakes, M. (1986). *Statistical inference: A commentary for the social and behavioral sciences.* New York: Wiley.

O'Connor, E. F. (1972). Extending classical test theory to the measurement of change. *Review of Educational Research, 42,* 73-97.

Ozer, D. J. (1985). Correlation and the coefficient of determination. *Psychological Bulletin, 97*, 307-315.

Page, R. (1990). Shyness and sociability: A dangerous combination for illicit substance use in adolescent males? *Adolescence, 25*(100): 803-806.

Pigott, T. D. (1994). Methods for handling missing data in research synthesis. In E. H. Cooper & L. V. Hedges (Eds.), *The handbook of research synthesis* (pp. 163-175). New York: Russell Sage Foundation.

Pillai, K. C. S. (1955). Some new test criteria in multivariate analysis. *Annals of Mathematical Statistics, 24*, 220-238.

Porter, A. C., & Raudenbush, S. W. (1987). Analysis of covariance: Its model and use in psychological research. *Journal of Counseling Psychology, 34*(4), 383-392.

Raudenbush, S. W. (1984). Magnitude of teacher expectancy effects on pupil IQ as a function of the credibility of expectancy induction: A synthesis of findings from 18 experiments. *Journal of Educational Psychology, 76*, 85-97.

Rawlings, J. O. (1988). *Applied regression analysis: A researcher's tool.* Belmont, CA: Wadsworth.

Ray, J. W., & Shadish, W. R. (1996). How interchangeable are different estimators of effect size? *Journal of Consulting and Clinical Psychology, 64*(6), 1316-1325.

Reed, J. F., & Slachert, W. (1981). Statistical proof in inconclusive "negative" trials. *Archives of Internal Medicine, 141*, 1307-1310.

Reichardt, C. S., & Gollob, H. F. (1987). Taking uncertainty into account when estimating effects. In M. M. Mark & R. L. Shotland (Eds.), *Multiple methods in program evaluation* (New Directions for Program Evaluation, 35, pp. 7-22). San Francisco: Jossey-Bass.

Rogosa, D. (1988). Myths about longitudinal research. In K. W. Schaie, R. T. Campbell, W. Meredith, & S. C. Rawlings (Eds.), *Methodological issues in aging research* (pp. 171-210). New York: Springer.

Rosenthal, R. (1995). Progress in clinical psychology: Is there any? *Clinical Psychology: Science and Practice, 2*(2), 133-150.

Rosenthal, R., & Jacobson, L. (1968). *Pygmalion in the classroom.* New York: Holt, Rinehart & Winston.

Rosenthal, R., & Rosnow, R. L. (1991). *Essentials of behavioral research: Methods and data analysis* (2nd ed.). New York: McGraw-Hill.

Rosenthal, R., & Rubin, D. (1986). Meta-analytic procedures for combining studies with multiple effect sizes. *Psychological Bulletin, 99*, 400-406.

Rosenthal, R., & Rubin, D. B. (1979). A note on percent variance explained as a measure of the importance of effects. *Journal of Applied Social Psychology, 9*, 395-396.

Rosenthal, R., & Rubin, D. B. (1982). A simple, general purpose display of magnitude of experimental effect. *Journal of Educational Psychology, 74*, 166-169.

Rosnow, R. L., & Rosenthal, R. (1989a). Definition and interpretation of interaction effects. *Psychological Bulletin, 105*, 143-146.

Rosnow, R. L., & Rosenthal, R. (1989b). Statistical procedures and the justification of knowledge in the social sciences. *American Psychologist, 44*, 1276-1284.

Rosnow, R. L., & Rosenthal, R. (1995). "Some things you learn aren't so": Cohen's Paradox, Asch's Paradigm, and the interpretation of interaction. *Psychological Science, 6*(1), 3-9.

Rosnow, R. L., & Rosenthal, R. (1996). Computing contrasts, effect sizes, and counternulls on other people's published data: General procedures for research consumers. *Psychological Methods, 1*(4), 331-340.

Rothman, K. (1986). Significance questing. *Annals of International Medicine, 105*, 445-447.

Roy, S. N. (1953). On a heuristic method of test construction and its use in multivariate analysis. *Annals of Mathematical Statistics, 26*, 117-121.

Rozeboom, W. W. (1960). The fallacy of the null hypothesis significance test. *Psychological Bulletin, 57*, 416-428.

Saunders, S. M., Howard, K. I., & Newman, F. L. (1988). Evaluating the clinical significance of treatment effects: Norms and normality. *Behavioral Assessment, 10*, 207-218.

Saville, D. J. (1990). Multiple comparison procedures: The practical solution. *The American Statistician, 44*, 174-180.

Scariano, S., & Davenport, J. (1987). The effects of violations of the independence assumption in the one way ANOVA. *The American Statistician, 41*, 123-129.

Scarr, S. (1997). Rules of evidence: A larger context for the statistical database. *Psychological Science, 8*(1), 16-17.

Schaffer, W. D. (1991). Power analysis in interpreting statistical nonsignificance. *Measurement and Evaluation in Counseling and Development, 23*, 146-148.

Schmidt, F. L. (1992). What do data really mean? Research findings, meta-analysis, and cumulative knowledge in psychology. *American Psychologist, 47*(10), 1173-1181.

Schmidt, F. L. (1996). Statistical significance testing and cumulative knowledge in psychology: Implications for training of researchers. *Psychological Methods, 1*(2), 115-129.

Schmidt, F. L., & Hunter, J. E. (1996). Measurement error in psychological research: Lessons from 26 research scenarios. *Psychological Methods, 1*, 199-223.

Schmitt, S. A. (1969). *Measuring uncertainty: An elementary introduction to Bayesian statistics.* Reading, MA: Addison-Wesley.

Sedlmeier, P., & Gigerenzer, G. (1989). Do studies of statistical power have an effect on the power of studies? *Psychological Bulletin, 105*, 39-316.

Serlin, R. C. (1993). Confidence intervals and the scientific method: A case for Holm on the range. *Journal of Experimental Education, 61*(4), 350-360.

Shadish, W. R., Hu, X., Glaser, R. R., Kownacki, R., & Wong, S. (1998). A method for exploring the effects of attrition in randomized experiments with dichotomous outcomes. *Psychological Methods, 3*(1), 3-22.

Share, D. L. (1984). Interpreting the output of multivariate analyses: A discussion of current approaches. *British Journal of Psychology, 75*, 349-362.

Shaver, J. (1993). What statistical significance testing is, and what it is not. *Journal of Experimental Education, 61*(4), 293-316.

Sheskin, D. J. (1997). *Handbook of parametric and nonparametric statistical procedures.* Boca Raton, FL: CRC Press.

Siegel, S. (1956). *Nonparametric statistics for the behavioral sciences.* New York: McGraw-Hill.

Smith, W., & Sasaki, M. S. (1979). Decreasing multicollinearity: A method for models with multiplicative functions. *Sociological Methods and Research, 8*, 35-56.

Snedecor, G. W., & Cochran, W. G. (1987). *Statistical methods* (8th ed.). Ames: Iowa State University Press.

Snyder, P., & Lawson, S. (1993). Evaluating results using corrected and uncorrected effect size estimates. *Journal of Experimental Education, 61*(4), 334-349.

Stevens, J. (1992). *Applied multivariate statistics for the social sciences* (2nd ed.). Hillsdale, NJ: Lawrence Erlbaum.

Stevens, S. S. (Ed.). (1966). *Handbook of experimental psychology.* New York: Wiley.

Stuart, A. (1954). The correlation between variate values and ranks in samples from a continuous distribution. *British Journal of Statistical Psychology, 7*, 37-44.

Summers, R. (1991). The influence of affirmative action on perceptions of a beneficiary's qualifications. *Journal of Applied Social Psychology, 21*(15), 1265-1276.

Tabachnick, B. G., & Fidell, L. S. (1996). *Using multivariate statistics* (3rd ed.). New York: HarperCollins.

Tatsuoka, M. M. (1971). *Significance tests.* Champaign, IL: Institute for Personality and Ability Testing.

Thompson, B. (1993). The use of statistical significance tests in research: Bootstrap and other alternatives. *Journal of Experimental Education, 61*(4), 361-377.

Thompson, B. (1994). The pivotal role of replication in psychological research: Empirically evaluating the replicability of sample results. *Journal of Personality, 62*(2), 157-176.

Thompson, B. (1997, August). *If statistical significance tests are broken/misused, what practices should supplement or replace them?* Paper presented at the meeting of the American Psychological Association, Chicago.

Titus, T. G. (1994). *The conceptualization and conduct of psychological research.* Unpublished manuscript, Spalding University, Louisville, KY.

Tucker, L., Damarin, F., & Messick, S. (1966). A base-free measure of change. *Psychometrika, 31*(4), 457-473.

Tukey, J. W. (1977). *Exploratory data analysis.* Reading, MA: Addison-Wesley.

Tukey, J. W. (1991). The philosophy of multiple comparisons. *Statistical Science, 6*, 100-116.

Underwood, B. J. (1957). Interference and forgetting. *Psychological Review, 64*, 49-60.

Velleman, P., & Hoaglin, D. C. (1981). *Applications, basics and computing of exploratory data analysis.* Boston: Duxbury, 1981.

Wainer, H. (1984). How to display data badly. *The American Statistician, 38*, 137-147.

Wampold, B. E., & Freund, R. D. (1987). Use of multiple regression in counseling psychology research: A flexible data-analytic strategy. *Journal of Counseling Psychology, 34*, 372-382.

Winer, B. J. (1971). *Statistical principles in experimental design.* New York: McGraw-Hill.

Wright, R. L. (1976). *Understanding statistics: An informal introduction for the behavioral sciences.* New York: Harcourt, Brace, Jovanovich.

Zimmerman, D. W., & Zumbo, B. D. (1989). A note on rank transformations and comparative power of the student *t* test and Wilcoxon-Mann Whitney test. *Perceptual and Motor Skills, 68*(3), 1139-1146.

Zwick, R. (1997, March). *Would the abolition of significance testing lead to better science?* Paper presented at the annual meeting of the American Educational Research Association, Chicago.

Index

About the Authors

Rae R. Newton is Professor of Sociology and Director of the Sociology Graduate Program at California State University, Fullerton. He is also Senior Statistician at the Center for Research on Child and Adolescent Mental Health Services, an NIMH-funded center at Children's Hospital in San Diego, California. He received his Ph.D. in sociology from the University of California, Santa Barbara, and completed postdoctoral training in mental health measurement at Indiana University. His interests include statistical methods in the social sciences, particularly structural equation modeling, mental health measurement, family violence, and graduate education. He is author, with Kjell Erik Rudestam, of *Surviving Your Dissertation: A Comprehensive Guide to Content and Process,* and of numerous articles in professional journals from a variety of disciplines.

Kjell Erik Rudestam is Associate Dean and a member of the faculty at The Fielding Institute in Santa Barbara, California. Previously, he was Professor of Psychology at York University in Toronto. He is a Fellow of the American Psychological Association (Division 12) and a Diplomate of the American Board of Examiners in Psychology (Clinical). He earned his Ph.D. degree in psychology from the University of Oregon. He is the author of three previous academic texts and joint author, with Rae R. Newton, of *Surviving Your Dissertation: A Comprehensive Guide to Content and Process.* In addition, he is the author of numerous articles in professional journals in his areas of interest, which include individual and systems change processes, crisis and suicide, and research methodology.

Statistical Analysis: An Overview

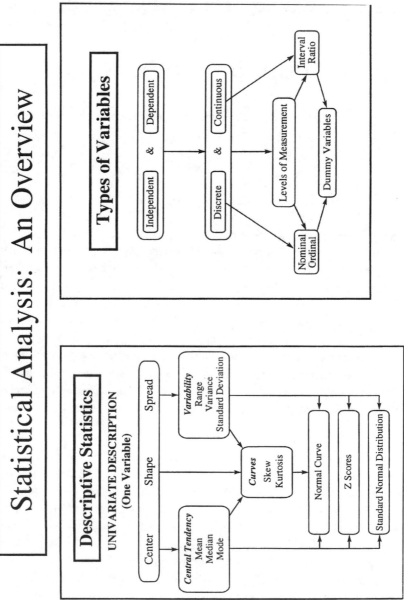